CYBER CRIME
DEFENSE GUIDE

A Plain English Guide for Protecting
Your Home, Network, Family, and Data

EDWARD GRASSIA

ROGUE TECH SERVICES

Cyber Crime Defense Guide: A Plain English Guide for Protecting Your Home, Network, Family, and Data

© 2025 Edward Grassia

ISBN: 979-8-9985877-0-2
First Edition
Published in the United States

For more information or to contact the author, please contact:
Rogue Tech Services LLC
5441 S Macadam Avenue, Suite R
Portland, OR 97239
Phone: 541-402-6357
roguetechservices.com

Dedicated to all the people who taught me.

Contents

Introduction

I BEGAN TINKERING WITH COMPUTERS MORE than thirty-five years ago, back when storage came on floppy disks and the internet was a far cry from what it is today. I'm talking slow dial-up and Bulletin Board Services (BBSs). Since then, I've witnessed and embraced every major shift: wireless, cloud computing, smartphones, virtualization, the Internet of Things (IoT), and more. This breadth of experience has shaped my perspective: technology, while empowering, demands constant vigilance.

For more than three decades, I've immersed myself in the evolving world of technology, from working on robots and computer-controlled manufacturing equipment to managing school district networks and defending them against cyber threats. While technology has undergone drastic changes in that time, the core principle remains the same: striking a balance between innovation and security.

After years of helping friends and family with IT issues, it became clear that there are very few, if any, sources written in plain language that non-technical people can use and understand. Technical manuals and documentation are often written by individuals who already have experience with the devices and software and aren't in a format that is easily understandable to tech newcomers.

I believe cybersecurity shouldn't be an enigma reserved for tech gurus or large corporations. It needs to be accessible and practical for everyday users, including parents, small business owners, retirees, and yes, even children.

My *Cyber Crime Defense Guide* is here to help you by focusing on straightforward measures for personal and home network security, and addressing specific scenarios such as remote work, social media, and children's online safety. This book reflects my years of experience distilled into actionable steps that anyone can follow.

The internet is a double-edged sword; it provides endless information, entertainment, and convenience, but it also introduces risks to personal information, finances, and physical safety. Many people believe that cybersecurity is too complicated and best left to professionals, but the reality is that even small steps can significantly reduce risk. This guide is designed for individuals with little-to-no experience in technology; those who aren't interested in becoming cybersecurity experts but want to secure their digital lives and protect their families from everyday online threats.

Rather than assuming you know how to configure advanced settings, I'll guide you through each step in plain language. Remember that every device is slightly different, so consult your specific product's manuals or online documentation if you encounter any issues. I intentionally avoid using brand names — other than a few such as Apple, Google, and Microsoft — and refer to browsers by their names, as there are dozens, if not hundreds, of different products in each category I cover. This book focuses on concepts and concrete steps you can take, not brand names and product-specific instructions.

My book is designed for anyone who wants to safeguard their household's digital environment, particularly those without a deep technical background. Whether you're a parent seeking to protect your children from malicious content, a small business owner working from a home office, or someone who wants to ensure your data is secure, this guide offers practical strategies and action plans. Each chapter will break down a core aspect of cybersecurity, covering everything from your home network and Internet of Things (IoT) devices to everyday activities such as password creation, app usage, and data backups.

I focus on the kinds of security measures that deliver the most significant impact with the least complexity. You won't find an in-depth analysis of advanced threat detection algorithms or complicated corporate network setups. Instead, you'll learn the importance of securing your Wi-Fi router, why unique passwords are essential, how to identify phishing emails, and why software updates (*yes, those annoying pop-ups*) are crucial. Each topic is addressed in detail, complete with step-by-step instructions that reference the types of generic settings and terminology you'll encounter on the devices in your home.

My approach combines clarity, real-world examples, and a healthy dose of practicality. I'm not here to overwhelm you with acronyms or impractical advice. I want you to make each step count, so by the time you reach the final chapter, you'll have leveled up your security without needing a second mortgage on your patience, sanity, or house.

As you work through this book, you'll make informed decisions about your security posture and take action to improve it. Whether it's deciding how to configure a new smart doorbell or knowing which suspicious email to delete immediately, you'll have the know-how to act decisively. It's time to move from feeling overwhelmed to feeling empowered. Let's begin securing your digital life one priority at a time.

Thank you for reading and for taking a stand in securing your digital life. I hope this book serves as a valuable reference, helping you navigate and enjoy the ever-changing tech landscape with confidence and peace of mind. ‡

How to Use This Book

THE BEST PLAN IS WORTHLESS IF you don't know how to apply it. Think of this book as a road trip: sure, you can skip around chapters, but a structured route ensures you won't miss the best attractions or crucial security measures along the way. I recommend reading each chapter sequentially if you're entirely new to this topic. However, feel free to jump around if you want to address a specific concern, such as setting up a new smart TV you've just bought. Each section stands independently, yet they all connect to provide a holistic approach.

KEY CONSIDERATIONS

- » **Modular design.** Each chapter can stand alone, for example, if you only want to learn about VPNs.
- » **Cumulative knowledge.** The chapters also build on each other, providing a cohesive, layered approach.
- » **Beating the dead horse.** You will see repetition and overlap as you read the book. This is designed to reinforce the critical topics and steps you can take to improve your security posture.

SELF-PACED LEARNING

- » **Adjust.** You might only have time to tackle one chapter a week. That's okay; security that fits your lifestyle will likely stick.
- » **Layer.** The best approach to security is to layer your defenses (there's that term again). Therefore, you will encounter references to topics you've already read about and those that will be discussed in later chapters.
- » **Organize.** Keep your notes or highlights accessible for quick lookups.

IMMEDIATE ACTIONS

> » **Scan the Table of Contents.** Note the chapters that address your immediate concerns; maybe "Identity Theft and Recovery" or "Children's Online Safety" are your priorities.

> » **Scan the Technical Terms.** Skim through this list of terms at the end of the book to familiarize yourself with some of the jargon you'll come across in the following chapters.

> » **Set learning goals.** Decide whether you will read cover-to-cover or select chapters in an order that aligns with your top priorities. Either way, commit to finishing. In security, the 80-percent-done approach can leave serious gaps.

> » **Bookmark and highlight.** Mark relevant sections or steps for quick reference. If a concept is confusing, leave a note to circle back after you've read subsequent chapters.

Keeping Pace with Evolving Threats

It's always important to keep in mind the broader landscape. Cyber threats evolve as rapidly as technology itself. New vulnerabilities surface every day, and attackers are quick to exploit them. Staying secure isn't just about setting a strong password once; it's about adopting a habit of vigilance and curiosity:

> » **Regular updates.** Learning to embrace (and not ignore) updates to your devices and applications.

> » **Threat awareness.** Recognizing that phishing techniques are constantly evolving and new forms of social engineering emerge daily. Especially now with the rise of AI tools.

> » **Continuous learning.** Monitoring news about significant data breaches or vulnerabilities can help you adjust your habits quickly when a new risk emerges.

Overcoming the Fear Factor

If you find the internet intimidating or worry about making a mistake, you're not alone. Many people get stuck in the mindset that they aren't "tech-savvy" enough to take control of their digital security. But remember that you don't need to be an expert to achieve a strong baseline of protection. Much like essential car maintenance, you only need a solid set of fundamentals: you know to change the oil, keep the tires inflated, and pay attention to warning lights in order to keep everything running smoothly.

Cybersecurity works similarly: by following a few best practices regularly, you'll be more secure than most users. This is also known as "layered security." *You will see me use this term often.*

Who Is This Guide For?

In addition to the solutions presented here, I recommend consulting your product manuals and official documentation, as each device, application, or service has unique requirements. Far too many products are on the market for one book to address step-by-step instructions for any specific brand. However, this book will guide you on what to look for and how to configure your devices or software.

KEY CONSIDERATIONS

» **Broad audience with varied needs.** From younger users navigating their first social media accounts to retirees connecting with family online, everyone encounters digital risks. Security solutions differ based on how you use technology. These range from advanced tasks to bare-minimum internet browsing.

» **No tech degree required.** I use plain language. If your eyes glaze over while reading complex jargon, you're in the right place. You'll see references to everyday life examples rather than obscure industry talk.

» **Practical over perfect.** There is no perfect security plan. I aim for realistic steps anyone can follow.

» **Real-life scenarios.** If it's not relative to everyday life, it's not in this book.

IMMEDIATE ACTIONS

» **Self-assessment.** Ask yourself: What devices do I use the most? Which ones hold my most sensitive data? Start there.

» **Share the approach.** If you have family, friends, or colleagues who are less tech-savvy, consider sharing the knowledge you gain from this book with them. Or treat them to their own copy. Encourage everyone to join you in adopting these best practices.

» **Set learning goals.** Determine which chapters or sections are most relevant to your situation, such as child safety or VPNs for remote work. Maybe it's encryption basics?

Final Notes

Regardless of your technical comfort level, remember these core truths:

» **Layers work best.** Combine strong passwords, multi-factor authentication, encryption, software updates, and awareness for absolute security.

» **Children and parents require extra care.** Schools and parents must collaborate to protect young users by implementing clear boundaries and providing age-appropriate guidance. Elderly parents may use the internet for both entertainment and necessity, but they may become overwhelmed by the complexities involved.

» **Stay curious.** Threats evolve rapidly. Keep learning, adapt your defenses, and share your knowledge with those who might be more vulnerable.

Thank you for reading and for taking a stand in securing your digital life. I hope this book serves as a valuable reference, helping you navigate and enjoy the ever-changing tech landscape with confidence and peace of mind. ⚡

CHAPTER ONE

The Triad of Cybersecurity

IN TODAY'S INTERCONNECTED WORLD, OUR LIVES revolve around digital technology more than ever. From smart thermostats that automatically adjust your home's temperature to online banking platforms that handle your most sensitive financial transactions, the internet has become the invisible thread weaving through our daily routines. But like all powerful tools, the internet comes with risks that, if left unaddressed, can have real-world consequences, both personal and financial.

For many, the term "cybersecurity" evokes images of hooded hackers typing furiously in dimly lit rooms or massive, complex systems that only experts can comprehend. This perception often leads to an assumption that cybersecurity practices are beyond the average person's reach, creating a dangerous sense of complacency. It's easy to think, "I'm just a regular user; I don't need to worry about hackers." However, malicious actors target everyday users precisely because they assume personal systems and accounts are insufficiently protected.

The good news is that cybersecurity doesn't have to be a burden. Even simple, incremental steps can significantly reduce your overall risk profile. In the same way you might lock your doors at night and close your windows when you leave home, there are straightforward measures you can take to protect your digital world. Think of this book as a step-by-step guide that translates cybersecurity jargon into plain English, arming you with practical knowledge you can apply immediately, with no advanced degree required.

The Expanding Digital Frontier

We are living in an era of ever-expanding digital transformation. Not so long ago, having a home computer was considered a luxury, and smartphones didn't exist. Today, a household might have multiple smartphones, tablets, laptops, smart speakers, TVs, security cameras, and even internet-connected cars, refrigerators, and light bulbs. Each of these devices represents a potential entry point for unauthorized access. The more devices you own and connect, the more critical it becomes to establish solid security practices.

Many of the solutions discussed in this book are either free or have a low cost. So you don't need to spend thousands of dollars and hundreds of hours securing your digital environment. However, you may need to do so if you don't protect yourself now. Once you've been breached or had your identity stolen, you'll spend money on repairs and protecting yourself against future attacks. However, it will be too late to recover what you have already lost, whatever that may be.

Why Everyday Users Are Prime Targets

It's essential to understand why cybercriminals often focus on individual users:

» **Volume and vulnerability.** There are far more individual home networks than corporate networks, and many are poorly secured. Attackers rely on the fact that many people use default passwords or outdated devices.

» **Valuable data.** Personal data — social security numbers, addresses, and credit card information — is what cybercriminals need to commit fraud or identity theft. The payoff can be huge for them, while the damage can be devastating for you.

» **Ease of exploitation.** Individuals are less likely to have dedicated IT teams, intrusion detection systems, or automated threat monitoring. Hackers know that a single unsecured device can open the door to an entire digital ecosystem, much like your house.

Consequences of a Breach

A security breach isn't just an abstract problem for big companies and government agencies. If your home network or personal accounts are compromised, the potential consequences can be immediate and severe:

» **Financial loss.** Attackers can gain unauthorized access to your bank accounts, open new accounts in your name, incur charges on your credit cards, or exploit your financial credentials in other ways.

» **Identity theft.** Once someone steals your personal information, it can take months or even years to recover fully, impacting your credit score and personal reputation.

» **Privacy invasion.** Hackers could access your private files, pictures, emails, or even live video feeds from cameras in your home, posing a serious threat to your privacy and safety.

» **Loss of trust.** When your friends, family, or colleagues receive scam messages or emails that appear to be from you, it can undermine their trust in your communications.

Confidentiality, Integrity, and Availability

If you've ever seen the acronym "CIA" floating around in cybersecurity discussions, we are not discussing *that* clandestine organization. Instead, CIA stands for **Confidentiality, Integrity, and Availability**, the foundational triad upon which modern cybersecurity rests. Think of these three elements as pillars supporting your network, accounts, and data. If one pillar is weakened, the whole structure wobbles.

I'll explore why these three concepts form the bedrock of security, how they're similar and different, and what practical methods keep each element strong. By the end, you'll see that "CIA" is less about shadowy espionage and more about building trust in your digital environment.

The Triad in Context

Most people think of "hacking" as outsiders stealing data, but data theft is just one piece of the puzzle. A robust security stance ensures that data remains confidential, maintains integrity, and is available when needed. Each part of the CIA triad safeguards a different angle of your digital life:

» **Confidentiality.** Prevents unauthorized disclosure of information.

» **Integrity.** Ensures data hasn't been altered and can be trusted.

» **Availability.** This guarantees you can use the data and systems when needed.

Whether you're a remote worker handling sensitive company data or a parent trying to protect your kids' online footprint, these three pillars affect every security decision. They also serve as a quick litmus test when deciding on new security measures. Ask yourself: *Does this protect confidentiality? Uphold integrity? Maintain availability?*

WHY THIS MATTERS

» **Core principles.** This triad is to cybersecurity what the ABCs are to language; the fundamental building blocks.

» **Unified focus.** Many people focus solely on "Confidentiality." However, ignoring Integrity or Availability can leave glaring gaps in your defenses.

» **Scalable approach.** The CIA triad is flexible, whether you're securing a family email account or protecting corporate trade secrets.

KEY CONSIDERATIONS

» **Balance.** Overemphasizing confidentiality can impede easy access (availability), leading to frustrated users circumventing security controls.

» **Integration.** The tools you deploy, encryption for confidentiality, checksums for integrity, and backups for availability should all work well together.

» **Policy and culture.** Technology alone can't do the job. Adopting a "security culture," *yes, even at home,* fosters consistent respect for the CIA triad at every level.

IMMEDIATE ACTIONS

» **Identify sensitive data.** Which documents, databases, or files require the highest confidentiality?

» **Check revision controls.** Do you track changes or maintain version histories for crucial files to preserve integrity?

» **Plan for uptime.** Do you have alternative methods to access critical systems if your primary network fails? Simple steps, such as offline backups or a secondary internet connection, can be lifesaving. Most people will not have a secondary internet connection for their house.

But this is where backups and offline files become essential if you lose connectivity for an extended period.

PUTTING IT ALL TOGETHER

All three pillars, Confidentiality, Integrity, and Availability, operate in synergy. Focusing on one at the expense of others can lead to bottlenecks, data corruption, or unauthorized access. By regularly reviewing your workflows, confirming backups, and testing encryption methods, you ensure that each element is adequately enforced.

The Three Pillars in Harmony

By understanding how each pillar supports the others, you can create a cohesive strategy that protects your data without compromising usability or accessibility.

WHY THIS MATTERS

» **Clear definitions.** Before applying tools or policies, it is essential to clearly understand each element.

» **Risk assessment.** Knowing precisely what each element entails lets you pinpoint vulnerabilities.

» **Security accountability.** Teams (or households) that understand the "why" behind each element tend to follow best practices more consistently.

KEY CONSIDERATIONS

» **Contextual relevance.** The approach to confidentiality in a corporate setting might be overkill for a casual home user and vice versa.

» **Tech lifespan.** Keep hardware and software up to date. An outdated router or operating system can compromise all three CIA elements simultaneously.

» **Human factor.** Even the best solutions crumble if you ignore or inadvertently override them, for example, writing your password on a sticky note.

IMMEDIATE ACTIONS

» **Define requirements.** Decide which data — medical records, financial files, family photos — demands maximum confidentiality, integrity, and availability.

» **Start small.** Implement basic encryption on personal files or use a backup solution. Build from there.

» **Communicate.** If multiple people share the same network, ensure everyone understands your security rules and the "why" behind them.

PUTTING IT ALL TOGETHER

The triad might sound academic, but it's surprisingly practical to break it down into daily habits, such as locking screens, creating backups, or verifying file authenticity. Think of the CIA triad as a cheat sheet for every security question:

» Does it protect privacy and Confidentiality?

» Does it maintain data Integrity?

» Does it remain accessible in terms of Availability?

Triad Parallels

Understanding each pillar's unique demands prevents a lopsided security approach. By mapping out where each pillar stands in your priority list, you can apply the right tools without stifling day-to-day usage or busting your budget.

» All protect data.

» All require oversight.

» All are non-negotiable.

WHY THIS MATTERS

» **Unified goals.** Although each pillar has a distinct focus, they share a common purpose: to shield and preserve data.

» **Baseline security posture.** Regardless of the simplicity or complexity of your tech setup, each pillar requires attention.

» **Layered defense.** Similar tools, such as firewalls, user authentication, and data monitoring, can enhance multiple CIA elements simultaneously.

KEY CONSIDERATIONS

> » **Universal relevance.** The triad applies to every environment, from personal devices to enterprise servers.
>
> » **Complementary toolsets.** Encryption primarily serves to maintain Confidentiality, but it also helps ensure Integrity by preventing unauthorized changes.
>
> » **Simplicity vs.** overengineering. The same approach can often check two or even all three boxes, like strong authentication controls to keep data confidential and integral.

IMMEDIATE ACTIONS

> » **Review current measures.** Which tools or processes address multiple elements? For instance, are your backups secured (Confidentiality) and verified (Integrity)?
>
> » **Look for overlaps.** A robust access control system enforces Confidentiality and can help protect Integrity.
>
> » **Streamline.** If you notice multiple solutions performing the same task, consolidate them to simplify the process.

PUTTING IT ALL TOGETHER

In many cases, a well-chosen security measure will boost more than one aspect of the CIA triad. Recognizing these overlaps saves time, money, and frustration. By viewing the triad holistically, you streamline your defenses without compromising thoroughness.

Balancing Security and Risk

Whether you're securing a home network or managing sensitive work data, finding the right balance ensures both protection and functionality.

WHY THIS MATTERS

> » **Tailored defenses.** A measure intended to protect confidentiality, such as encryption, may do little to enhance Availability if backups or redundant systems are necessary.
>
> » **Priority shifts.** In a hospital, Availability can take precedence; imagine

how life-critical data must be accessible at all times. But in a spy agency, Confidentiality may take precedence over all else.

» **Risk management.** Knowing the distinct vulnerabilities of each pillar ensures that your overall plan is not flawed.

KEY CONSIDERATIONS

» **Potential conflicts.** Security policies that tightly lock data might hinder Availability or day-to-day operations.

» **Cost benefit.** Integrity checks can be resource-intensive; therefore, weigh the criticality of the data before implementing advanced solutions.

» **Incident response.** A breach of Confidentiality vs. a lack of Availability triggers different crisis protocols.

IMMEDIATE ACTIONS

» **Prioritize.** Identify which element is most critical in each scenario or device, for example, 24/7 server uptime versus the Confidentiality of medical forms.

» **Customize.** Don't copy and paste security measures across all devices. Tailor them based on the function and value of the data involved.

» **Monitor continuously.** Distinct warning signs exist for each pillar. For example, a Confidentiality breach might manifest as unauthorized access logs, while Availability issues manifest as repeated downtime.

PUTTING IT ALL TOGETHER

Understanding each pillar's unique demands prevents a lopsided security approach. By mapping out where each pillar stands in your priority list, you can apply the right tools without stifling day-to-day usage or busting your budget.

The SUC Triangle

I have had conversations about another similar triad over the years. Called the SUC triangle, it stands for "Security," "Usability," and "Cost." The theory is that moving one of those items up or down will influence the others. For example, if you raise Security, the cost is likely to increase, and Usability will decrease. Why?

Added Security typically comes from deploying software or hardware tools, many of which incur Costs. Those tools also add layers of complexity to an environment, which influences the ease of Usability for you and other people on your network.

Conversely, if you improve Usability, you will likely have to reduce Security, which will also lower your Cost. However, you are then taking on additional risk.

The moral of the story is that security is a balancing act between addressing risk and vulnerabilities while still allowing you to use your email or stream a movie.

Tools and Methods

Let's look at some tools and methods used to implement and manage each element in the CIA triad.

CONFIDENTIALITY

» **Encryption.** File-level, disk-level, communications.

» **Access controls.** Multi-factor authentication; least-privilege access.

» **Physical security.** Locked server rooms; screen locks.

INTEGRITY

» **Checksums and hashes.** Used to verify data hasn't changed.

» **Version control.** Ability to restore files to previous states.

» **Digital signatures.** Guarantee authenticity and non-repudiation.

AVAILABILITY

» **Redundant systems.** Hot backups; multiple data centers.

» **Regular backups.** Onsite or offsite; automatic scheduling.

» **Failover strategies.** Battery backups; secondary internet lines.

WHY THIS MATTERS

» **Actionable steps.** Talking theory is easy; implementing solutions is what truly secures your systems.

» **Futureproofing.** The tools that protect each pillar also improve as threats evolve. It is essential to monitor updates.

» **Budget and ROI.** Some solutions are pricey or resource-heavy; choose the ones that give the best security return for your needs.

KEY CONSIDERATIONS

» **Compatibility.** Ensure the tools you choose integrate smoothly. For example, your encryption solution shouldn't compromise system availability.

» **Regular maintenance.** Tools are not "set and forget." They require updates, monitoring, and occasional replacement.

» **User training.** Even the best technology fails if you don't follow guidelines, ignore updates, or disable "inconvenient" features.

IMMEDIATE ACTIONS

» **Tool audit.** List the security tools you currently use. Note which pillar(s) they support.

» **Gap analysis.** Identify missing measures. Maybe you have encryption (Confidentiality) but no checksums for Integrity.

» **Roadmap.** Prioritize which tools or methods to implement next based on the criticality of your data and your budget.

PUTTING IT ALL TOGETHER

One of the biggest challenges in cybersecurity is the ever-changing nature of threats, which means no single solution is a permanent fix. Instead, it's a continuous process of auditing your devices and accounts, learning about criminals' new tactics, and adjusting your defenses accordingly. This book won't turn you into a cybersecurity expert, but it will provide you with the practical tools you need to build a stable and secure foundation. Specific tools, technologies, and procedures map neatly to each corner of the triad. By selecting the right combo for your unique environment, you ensure that Confidentiality, Integrity, and Availability work harmoniously rather than stepping on each other's toes.

The "CIA" triad, comprising Confidentiality, Integrity, and Availability, serves as your guiding compass for cybersecurity. Each part plays a distinct role, yet they all share the same objective: protecting data from unauthorized exposure, tampering, and disruptions. Recognizing their similarities helps you see how a single solution can defend multiple pillars; understanding their differences allows you to focus your resources more effectively.

Remember: no single product or policy magically secures everything. Instead, combine the correct methods for each scenario, like encryption for confidentiality, hashing for integrity, and backups for availability. Applying the CIA triad consistently and thoughtfully will create a stronger, more resilient foundation for all your digital defenses. That means fewer sleepless nights, fewer frantic phone calls to tech support, and a far better shield against ever-evolving threats.

You've learned how the CIA triad underpins cybersecurity philosophy. In the following chapters, I'll delve into practical topics, from antivirus software to identity theft prevention, each reinforcing the same fundamental principle: protect the data, keep it accurate, and ensure that the right people can access it when needed. That's the fundamental mission of the CIA triad.

Every victory — changing a weak password to a robust one or installing a firmware update — adds another layer of protection around you and your family. Along the way, you'll discover that your confidence in technology grows in tandem with your increasing security skills. The steps I cover may seem small, but collectively, they make a difference in protecting your digital life. *It's all about the layers.* ‡

CHAPTER TWO

Prioritizing Security

I'M SURE YOU ARE THINKING, *WHERE do I start?* It's easy to feel paralyzed by the sheer volume of security advice floating around. Digital threats come in all shapes and sizes, from malware lurking in a shady app to hackers scanning for unsecured home networks. Every day, you might see a new headline about some vulnerability or data breach. If you're wondering where to begin, don't worry. The path to strong cybersecurity starts with a handful of foundational tasks that deliver the highest return on your security investment. In other words, focus on what provides the most significant impact first rather than trying to address every single threat at once.

Below, I'll explore five priority areas you can tackle immediately to establish a solid security baseline. Think of these steps as the pillars of your digital fortress: if one is weak, the rest are at risk. By systematically addressing each, you'll quickly enhance your overall security posture and significantly reduce the likelihood of falling victim to common attacks. I dive deeper into these topics in further chapters, but the importance of these topics can't be overlooked.

The Weakest Link

You probably figured out I'm not talking about the game show! There are many conversations and differing opinions in the security world about the weakest link in security. Is it the tools, a lack of standards, lax attitudes, a lack of knowledge, or the users?

I'm not here to convince you of which one it is. After completing this book, you will be able to draw your own conclusions. But I will state a fact that can't be ignored as you travel down the road to a secure environment. No single combination of settings and tools will guarantee your complete safety. *You*, along with anyone else who has access to your systems, are essential to establishing and maintaining your security.

SPOILER ALERT: Security is an ongoing job as long as you use technology.

Even if you complete every task in this book, you must continue to maintain your hard work. If you let things slide and become outdated, you'll have to start from scratch. I'm not saying this is a full-time job for the rest of your life. But like changing the oil in your car and rotating the tires, it is critical to keep up the maintenance of your digital life.

Tame the Overwhelming: A Strategic Approach

Many new to cybersecurity find themselves either trying to do everything at once or giving up and doing nothing. The sweet spot is to do a few critical things well and then build upon that foundation as your knowledge and confidence grow.

I'll cover the most common entry points attackers exploit by focusing on high-impact steps, such as securing your home Wi-Fi network, adopting strong passwords, enabling multi-factor authentication, using antivirus software, keeping software up to date, and learning to spot phishing scams. You'll already be ahead of most casual cybercriminals with these core defenses.

Priority One: Secure Your Home Wi-Fi Network

Your home Wi-Fi is the gateway to everything else in your digital life. If compromised, attackers can access your computers, smart devices, and personal data across the network.

WHY THIS MATTERS

A poorly secured or unsecured Wi-Fi network is like leaving your front door unlocked. Hackers routinely scan for unprotected or weakly protected wireless networks, knowing they can often enter without being detected.

Securing your Wi-Fi network is the single most critical step you can take right now, and it lays the groundwork for everything else in this guide.

KEY CONSIDERATIONS

» **Router configuration.** Change your router's default administrator password. The factory defaults are publicly known and can be easily searched online.

» **Encryption protocols.** Choose WPA2 or WPA3 (if supported) to ensure your data is encrypted in transit. If your router is outdated and doesn't support the latest wireless standards, consider replacing it.

» **Guest networks.** Set up a dedicated guest network to keep visitors from accessing your primary devices. This also helps contain potential threats if someone's device is infected.

IMMEDIATE ACTION ITEMS

» **Log in to your router settings.** Find your router's IP address. This is usually listed in the manual or on a sticker. Enter it into your web browser's address bar, for example, where you would type amazom.com. The IP address will look something like this: https://192.168.168.168, but your numbers will most likely be different.

» **Change default credentials.** Replace the default credentials, such as "admin/admin" or similar weak passwords, with a strong, unique password. If it's an option, you should also consider changing your username.

» **Enable WPA2 or WPA3.** Navigate to the "Wireless Security" settings and select the highest encryption level your router supports.

» **Turn off remote management.** If it's on by default, disable this feature to prevent unauthorized remote access.

Priority Two: Use Strong, Unique Passwords

Passwords are often dismissed as a nuisance; We've all seen people scribbling a single password on a sticky note and using it across a dozen accounts. That practice is a hacker's dream come true.

A strong, unique password is your digital key. Just like you wouldn't use the same physical key for your house, car, and office, don't rely on a single password for multiple online services.

WHY THIS MATTERS

If a cybercriminal manages to crack or steal a single password you've reused on multiple sites, all those accounts become vulnerable. Weak or repeated passwords undermine every other security measure you put in place.

KEY CONSIDERATIONS

» **Password complexity.** Longer is stronger. Use a mix of uppercase and lowercase letters, numbers, and symbols.

» **Uniqueness.** One account = one unique password. Never use the same password on more than one account.

» **Password managers.** Tools that store and generate passwords can drastically reduce the mental load of remembering countless credentials. They reduce the burden of creating, using, and remembering a single password for every account. The more accounts you have, the more you will appreciate using a password manager.

IMMEDIATE ACTIONS

» **Identify critical accounts.** Begin with your email, banking, and social media accounts.

» **Change weak passwords.** Replace anything short or obvious (like "12345" or "password1"). You should create the longest and most complex password that a site or application will allow.

» **Avoid patterns.** Don't simply increment numbers or use easily guessed information, such as names or birthdays. If you have or use a password manager, utilize the password generator that comes with it to create complex and unique passwords for all the accounts you log into.

» **Implement a password manager.** This securely stores your credentials and can even auto-fill login pages.

Priority Three: Enable 2FA or MFA

Even the strongest passwords can be stolen through phishing, data breaches, or malware. Two-Factor Authentication (2FA) or Multi-Factor Authentication (MFA) adds an extra hurdle for hackers by requiring a second form of verification, often a temporary code or biometric check. Including 2FA/MFA in the login process now involves something you know (a username and password) and something you have (such as your phone, a security key, or an authenticator app).

Authenticator apps have gained preference over SMS texts since those can be intercepted, and the apps are free. Whichever route you choose, something is better than nothing. This is part of layering security *(are you keeping count?)* since every additional step or measure you add requires a hacker to jump through an extra hoop to breach your network or account.

WHY THIS MATTERS

Imagine someone gets ahold of your password. Without 2FA/MFA, they can log in instantly as you. With 2FA/MFA, they still need a one-time code sent to your phone or generated by an authenticator app. This significantly reduces the ease of unauthorized access.

Two-factor authentication (2FA) or multi-factor authentication (MFA) is one of the most effective tools against account hijacking. If there's a single piece of advice you take from this entire chapter, let it be to enable 2FA for your most critical accounts as soon as possible. Again, I recommend using it for every account that supports it.

KEY CONSIDERATIONS

» **Authentication methods.** These include SMS codes, authenticator apps, hardware keys, and biometric checks (such as fingerprints). Each has pros and cons, but any 2FA is better than none.

» **Where to enable.** Email accounts, social media, shopping sites, insurance sites, banking apps, your home router, and the like. *I recommend using it wherever it's available.*

» **Setup process.** This is typically located in the "Account Settings" section, under "Security" or "Login Options."

IMMEDIATE ACTION ITEMS

» **Check account settings.** Look for a "Security" or "Privacy" section and enable 2FA where possible.

» **Choose a 2FA method.** If available, use an authenticator app rather than SMS, as text messages can be intercepted or hijacked through SIM swaps.

» **Store backup codes safely.** Some services provide backup codes in case you lose access to your phone. Keep these codes secure, such as in a password manager or on a USB flash drive that is not connected to your network or devices.

Priority Four: Update Your Devices and Software

We've all been guilty of ignoring those annoying "Software Update Available" prompts. Unfortunately, each missed update allows criminals an opportunity to exploit known vulnerabilities. Keeping your software current is like closing an open window as soon as you find it, waiting only makes it easier for intruders to sneak in.

WHY THIS MATTERS

Software updates often contain security patches, fixes for flaws that attackers already know how to exploit. Delaying or skipping updates leaves you unnecessarily exposed.

KEY CONSIDERATIONS

» **Automatic updates.** Many operating systems (OS) and apps can update themselves if enabled. If you know you won't remember to do these manually, you should consider enabling automatic updates.

» **Firmware updates.** Routers, printers, and IoT devices also receive updates. These can be software, feature, or critical security updates. Some devices may require you to check for updates manually. If your device has an auto-update feature, enable it.

» **End-of-life products.** Devices or software no longer supported by the manufacturer should be replaced, as they won't receive essential security patches. Hackers are aware of this, too, and they will search for older, vulnerable devices connected to the internet. Websites like shodan.io enable anyone in the world to search for and locate your outdated devices. Don't be the one they find.

IMMEDIATE ACTION ITEMS

» **Enable automatic updates.** To turn this on, check the settings on your phone, computer, networked devices, and critical applications, such as Microsoft Office, Adobe, and internet browsers.

» **Schedule manual checks.** Pick a day each month to manually verify that all your devices have the latest security patches.

» **Replace unsupported devices.** If your device no longer receives updates, consider upgrading to a newer model.

» **Back up before updating.** For critical systems, perform a quick backup before installing major updates to prevent data loss in the event of an issue.

Priority Five: Recognize Phishing Attempts

Phishing is one of the oldest tricks in the hacker's playbook, yet it remains remarkably effective. Attackers send fraudulent emails or messages designed to steal or trick you into disclosing your login credentials, financial information, or personal details.

A phishing email may appear legitimate at first glance, but taking a moment to double-check can save you from significant trouble. The introduction of AI has only aided hackers in this realm, requiring users to be even more cautious. Malicious actors utilize AI to create realistic-looking fake videos, voice recordings, text messages, and phishing emails that target unsuspecting users. This may sound pessimistic, but assuming everything is fake until you verify it will reduce your risk of being fooled.

WHY THIS MATTERS

You can have robust passwords and up-to-date software, but if you inadvertently hand your login details to a scammer, all that effort is wasted. Phishing bypasses many technical defenses by exploiting human trust and inattention.

KEY CONSIDERATIONS

» **Look for red flags.** Poor grammar, urgent demands, payment requests, suspicious links, or unfamiliar sender addresses.

» **Verify before you click.** Hover over links — DON'T CLICK — with your mouse pointer to check the web address, and never download attachments from unknown or unverified sources.

» **No legitimate entity asks for passwords via email.** If someone claims to be from your bank or a government agency and demands immediate action, call them at a known official number to verify.

IMMEDIATE ACTIONS

» **Educate yourself and your family.** Ensure everyone understands that clicking on unfamiliar links can lead to malware or data theft.

» **Use spam filters.** Most email services have built-in filtering systems; adjust the settings for maximum efficiency.

» **Report suspicious emails.** If your email service supports it, mark suspicious messages as "phishing" or "spam" to help protect others.

» **Stay skeptical.** Adopt a healthy dose of skepticism whenever an email or text claims to have "urgent" news about your account.

PUTTING IT ALL TOGETHER

By tackling these priorities — securing your Wi-Fi, using strong passwords, enabling two-factor authentication (2FA), keeping software updated, and recognizing phishing — you'll address the most common avenues that cybercriminals exploit. Remember "Layered Security" *(there it is again)*: each layer strengthens your digital defense.

Once you've completed these essential tasks, you'll be ready to delve deeper into advanced topics, such as IoT device security or cross-platform best practices. However, even if you stop here, you've already taken a significant leap from being an easy target to becoming a savvy, security-minded user.

A Roadmap, Not a Race

Don't feel pressured to do everything in a single day. While each area is urgent, doing them properly is also critical. You don't want a false sense of security when you're done; you want actual improvements in your security. Start with the easiest one for you, such as changing your Wi-Fi password, and move on. As you gain confidence, you'll find that adopting secure habits will become second nature, and you'll be less likely to fall victim to the latest cyber threat.

In the following chapters, I'll break down the essentials of each priority and walk you through exactly how to execute them. By the time you reach the end of this book, you'll have the skills to maintain and evolve your security measures as technology and threats change.

So, take a deep breath, roll up your sleeves, and start securing your digital environment one priority at a time. ⁑

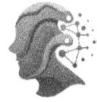

CHAPTER THREE

Securing Your Home Network

IN PERSONAL CYBERSECURITY, YOUR HOME NETWORK is the castle wall between your private information and the internet. It's where all your digital traffic, streaming services, email, remote work sessions, and smart home commands converge before entering or leaving your household. If cybercriminals breach this perimeter, they can access your data, compromise your devices, or even exploit your network for illicit activities. While that might sound dramatic, the good news is that securing your home network is one of the most straightforward steps you can take to protect yourself and your family. This chapter will show you exactly how to do it.

Build a Strong Home Network Defense

Securing your home network reduces the likelihood of a breach at this foundational level. The following steps focus on practical, high-impact measures that you can apply quickly, even if you don't consider yourself tech-savvy.

But before I dive into the practical steps, it's helpful to understand why home network security is so essential:

» **Single point of entry.** Much like the front door to your house, your home Wi-Fi router is the primary gateway to every connected device. Laptops, smartphones, tablets, smart cameras, and even connected light bulbs depend on the router to communicate with the internet. A single oversight in your router settings can expose everything else inside your network.

» **Common target.** Home networks are notoriously attractive to cybercriminals because many users leave default settings in place. Hackers often conduct automated scans for networks that still use factory passwords or outdated encryption, making them low-hanging fruit for malicious access.

» **Data flow.** Sensitive information, such as credit card details, passwords, and private emails, travels through your router. If it's compromised or uses weak encryption, eavesdroppers may read, intercept, or manipulate your data at will.

Change Your Router's Default Admin Password

Your first defense is securing the router's administrative interface, the control panel through which you manage network settings.

Taking five minutes to customize this one setting can foil many automated attacks that rely on default credentials. If your ISP allows it, you should also do this on the router they gave you. If you have a wireless router from your ISP and one that you own, you should turn off wireless on the one you don't use. Having an unmonitored, wide-open wireless network is just asking for trouble.

WHY THIS MATTERS

Routers typically come with default usernames and passwords, such as "admin" or "password," which are both well-known to cybercriminals and easily found online. If you never change these credentials, you're leaving the network's front door wide open.

STEP-BY-STEP GUIDE

» **Find your router's IP address.** This is often printed on the router or listed in its quick-start guide. Standard addresses include 192.168.0.1 or 192.168.1.1.

» **Look it up.** Type your router's address into your web browser's URL bar.

» **Log in using default credentials.** Check the router manual or the label on the device for the factory-set username and password. If you can't find them, search the manufacturer's online documentation.

» **Set a new administrator password.** Look for a settings tab, often labeled "Administration," "Maintenance," or "Security."

» **Create a strong, unique password.** This should include a combination of upper and lowercase letters, numbers, and special characters.

» **Confirm changes.** Save or confirm your settings as required.

» **Store your new credentials securely.** Ideally, you should use a password manager. Again, a password manager allows for unique, long, and complex passwords you don't have to remember or write down.

BEST PRACTICES

» **Make your password unique.** Don't reuse passwords you use elsewhere, such as your email or social media accounts.

» **Avoid common phrases.** Hackers can quickly guess passwords that contain essential dictionary words or personal information, such as birthdays or anniversaries.

» **Change it periodically.** Consider updating your router's admin password at least once a year or whenever you suspect a security issue.

Use WPA3 or WPA2 Encryption

Encryption ensures that data traveling over your Wi-Fi is scrambled and unreadable to anyone attempting to intercept it. Without proper encryption, a cybercriminal can intercept or inject malicious content into your network traffic.

Remember that many internet-connected "smart" devices, such as thermostats, garage door openers, and major appliances, do not use the latest wireless protocols. This may prevent you from using the latest technologies on your network and potentially create a vulnerability you can't avoid. When introducing these items into your home network, check the specifications to decide if it's worth the risk. Aside from lacking proper encryption, those devices may force you to use a slower connection if they don't support 5 or 6 GHz frequencies.

Encrypting your wireless signal might sound technical, but most modern routers make it straightforward. The result is a robust layer of privacy and security for your home network traffic.

WHY THIS MATTERS

Strong wireless encryption standards, like WPA3 or WPA2, prevent unauthorized users from connecting to your network without the correct passphrase. Using widely available tools, weaker standards such as WEP can be cracked in minutes.

STEP-BY-STEP GUIDE

» **Access your router settings.** Log back into your router's administration panel using the new credentials you just set. Look for a menu titled "Wireless," "Wi-Fi," or "Security."

» **Select the highest encryption standard.** If your router supports WPA3, choose that first. Otherwise, opt for WPA2 (AES). Avoid WPA/WPA2 mixed-mode if WPA2-only is available, as mixed-mode can sometimes weaken security.

» **Create a strong Wi-Fi passphrase.** Consider a passphrase that is long, complex, and not easily guessable. For example, "C0ff33BeansAre$tr0ng!" The goal is to make it secure and memorable — passphrases often work better than single words.

» **Apply changes and reconnect devices.** Click "Save" or "Apply." Your router may restart.

» **Reconnect.** You must reconnect all Wi-Fi devices on your network using the updated password.

BEST PRACTICES

» **Future-proof with WPA3.** If your current router doesn't support WPA3, consider it when buying your next one.

» **Avoid WEP.** WEP is so outdated that it offers no protection.

» **Separate personal and guest networks.** I'll cover this more later, but consider setting up a separate Wi-Fi network with a unique passphrase for guests and IoT devices.

Turn Off Remote Management

Many routers include a feature called "Remote Management" or "Remote Administration," which enables administrators to modify settings remotely outside the local network. While occasionally useful for troubleshooting, this introduces potential risk.

Disabling remote administration significantly reduces the likelihood that individuals with malicious intentions can compromise your router settings.

WHY THIS MATTERS

If remote management is enabled, cybercriminals who discover your router's IP address might gain direct access to its configuration interface, even from halfway around the world. This is especially dangerous if they manage to guess your router's credentials.

STEP-BY-STEP GUIDE

» **Locate remote management settings.** In your router's admin panel, search for advanced settings or a tab labeled "Remote Management" or "Remote Administration."

» **Disable the feature.** Uncheck the box or toggle it off to allow administration outside the local network.

» **Save changes.** Apply the updated configuration.

» **Verify it's off.** After your router restarts, recheck the remote management setting to ensure it stays disabled. Hardware can seem very temperamental at times. Verifying every change you make is a good practice to ensure it was applied correctly.

BEST PRACTICES

» **Use local management.** Make changes while connected to your local Wi-Fi or through an Ethernet cable.

» **Enable VPN instead.** If you genuinely need remote access, consider using a Virtual Private Network (VPN) that securely tunnels into your home network. Newer routers often come with a VPN feature.

Set Up a Guest Network

Your primary Wi-Fi network often hosts critical devices, including personal computers, mobile devices, and possibly even home office equipment. Providing visitors or peripheral devices (like IoT gadgets) direct access to this network increases the risk of a security breach. Implementing a guest network keeps untrusted or rarely monitored devices at a distance from your most sensitive data.

WHY THIS MATTERS

A guest network acts like a separate "tunnel" for all non-critical devices or visitors. If one of them is compromised, the attackers don't gain access to everything else on your primary network.

STEP-BY-STEP GUIDE

» **Enable guest Wi-Fi.** Within your router's wireless settings, look for an option to create or enable a guest network.

» **Choose a different SSID.** The SSID, or Service Set Identifier, is the name of your wireless network. Give the guest network a unique but easily identifiable name, such as "SmithGuestWiFi."

» **Protect with a password.** Don't leave your guest network open (unprotected). If possible, use the same WPA2 or WPA3 standard.

» **Isolate guest traffic.** Some routers offer a feature called "Client Isolation" or "Guest Isolation." This prevents devices on the guest network from seeing or interacting with your primary network devices or with each other.

» **Save your settings.** Reboot your router if necessary.

BEST PRACTICES

» **Share wisely.** Provide the guest network passphrase only to those who need it and update it periodically.

» **Avoid device overload.** If you're connecting numerous IoT devices, consider whether they'd be better on the guest network, isolated from your computers and smartphones.

» **Don't create a guest network if you won't use it.** It will be one less thing to manage.

Monitor Connected Devices

After taking the above steps, it's crucial to remain vigilant. Hackers constantly evolve their techniques; even the strongest passwords can be compromised in a breach. Regularly reviewing which devices are connected to your Wi-Fi can alert you to suspicious activity.

WHY THIS MATTERS

If an unfamiliar device appears in your router's device list, it could be a sign that someone is piggybacking on your network without permission. Prompt detection can allow you to boot them out before they cause any damage. Regular monitoring is akin to conducting a quick security patrol of your network to ensure everything is as it should be.

STEP-BY-STEP GUIDE

» **Log in to your router periodically.** Make it a habit, maybe once a month or whenever you suspect something is off.

» **Access the "Connected Devices" section.** This may be listed under "Device List," "DHCP Clients," or a similar category.

» **Check device names or MAC addresses.** Some devices automatically identify themselves by name ("John's iPhone"), while others display only a MAC address. Consult your device documentation to match addresses with known gadgets. You may be able to rename the device in its settings so that it appears on the network using the name you provided rather than a generic, unknown name.

» **Kick out intruders.** If you see something you don't recognize, most routers let you remove or block that device.

» **Change your Wi-Fi password.** If you discover any unknown connections, immediately update your Wi-Fi and router admin passwords to prevent re-entry.

BEST PRACTICES

» **Stay organized.** Keep a simple record of your family's devices so you can quickly identify a newcomer.

» **Set alerts (if available).** Some modern routers or companion mobile apps allow you to receive notifications when a new device joins.

ADDITIONAL CONSIDERATIONS

» **Router location.** The physical placement of your router can affect who can pick up its signal. Ideally, you should center it in your home to provide even coverage without broadcasting too far outside your walls.

» **Firmware updates.** Like any piece of technology, routers can have bugs or security vulnerabilities. Check periodically for firmware updates from the manufacturer and apply them as soon as they're available.

» **Replace outdated hardware.** Consider upgrading if your router is several years old and doesn't support modern encryption standards, such as WPA3. The investment is relatively small compared to the value of enhanced security and better performance.

» **Router firewall.** Many routers have a built-in firewall. Consult your router's manual or online documentation to ensure it is enabled and configured correctly. Turn off or disable any services or ports that you aren't using.

VPNs

In an increasingly connected world, privacy can feel like a luxury. That's where a **Virtual Private Network (VPN)** comes into play, cloaking your internet traffic in a secure tunnel and making it more challenging for snoops or malicious actors to track your activities. If you imagine the internet as a pipe through which everything flows, a VPN is a pipe inside that pipe, insulating your data from everyone else's.

But VPNs aren't just for anonymity buffs or corporate employees anymore. Whether you're shopping online, working remotely, or simply checking email at a coffee shop, a good VPN can help protect your data from prying eyes.

Think of internet traffic like cars driving down a highway: when the road is open, anyone can see which cars are passing by and where they're headed. But once those cars enter a tunnel, they're hidden from view.

I'll break down the fundamentals of VPNs, what they are, whether you need one, how to choose one, and why inbound VPN connections pose unique risks. Finally, I'll examine which devices should run a VPN client to keep your digital life private and secure.

VPN Considerations

A VPN helps reduce external surveillance and shields sensitive data, such as passwords or financial information, when using untrusted networks like public Wi-Fi.

However, like most security tools, VPNs aren't magic. They can slow your connection, and if you pick a dubious service, they might even sell or misuse your browsing data. It's crucial to balance privacy needs with performance and reliability.

WHY THIS MATTERS

» **Privacy and security.** A VPN encrypts your connection, preventing eavesdroppers on public Wi-Fi or untrusted networks from stealing your info.

» **Geographic flexibility.** It can appear you're browsing from a different location, bypassing geo-restrictions on content.

» **Business use.** Companies utilize VPNs to enable remote workers to access their internal networks securely.

KEY CONSIDERATIONS

» **Encryption tunnel.** Your data travels through an encrypted "tunnel," making it much harder for outsiders to read.

» **VPN protocols.** Common protocols include OpenVPN, WireGuard, IKEv2/IPsec, L2TP, and PPTP, each offering different levels of speed, security, and reliability.

» **Logging policies.** Some VPNs keep detailed logs of your activities, which defeats the purpose if you want absolute anonymity.

IMMEDIATE ACTIONS

» **Identify your needs.** Are you primarily protecting yourself on public Wi-Fi, or do you also want to bypass geo-blocks?

» **Check device compatibility.** Ensure the VPN you choose supports all your primary devices, including phones, tablets, and laptops.

» **Learn basic terms.** Understand the distinction between encryption protocols and logging policies, as these factors are crucial when selecting a provider.

PUTTING IT ALL TOGETHER

At its core, a VPN is a secure, encrypted channel between your device and the internet. Once you understand this principle, it becomes easier to see how it can benefit both casual users, say someone safe browsing at coffee shops, and power users, such as people accessing corporate servers from anywhere.

Do I Need a VPN?

If you're streaming Netflix and shopping on Amazon from your living room, it's probably not critical for you to have one. But if you value your privacy, handle sensitive information — whether personal or work-related — or simply prefer not to be tracked online, get a good VPN and use it consistently.

WHY THIS MATTERS

» **Threat environment.** Cybercriminals frequently target unsecured public Wi-Fi hotspots, such as airports and cafes, for data theft. A VPN minimizes this risk.

» **Privacy levels.** Your ISP can observe or log your internet traffic at home. A VPN can reduce that.

» **Alternative solutions.** Some people believe using only HTTPS or secure websites is enough to protect you. HTTPS only protects the connection between your browser and the website if the site supports it. A VPN protects all the data that passes through the connection.

KEY CONSIDERATIONS

» **Frequency of use.** If you rarely use public Wi-Fi, you may not need a VPN every day, but it's still nice to have when traveling. Even if you only use Wi-Fi at friends' or family's houses, since you can't guarantee the security of their network

» **Performance hit.** VPNs can slow your connection. The slowdown might be noticeable if you have limited bandwidth or older hardware.

» **Legal notices and terms of service.** Some websites and streaming services frown upon VPN usage. This could cause access or account issues.

IMMEDIATE ACTIONS

» **Evaluate your routine.** A VPN is highly recommended if you often find yourself on unfamiliar Wi-Fi networks.

» **Check work requirements.** Some companies mandate VPN usage for remote access or confidential tasks.

» **Consider your privacy priorities.** A VPN can be a game-changer if anonymity or privacy is a top priority.

» Putting It All Together

VPNs are not mandatory, but they provide an extra layer of protection against digital threats and surveillance. Consider your habits, risk level, and performance needs to decide if it's worth the added complexity and potential cost.

How Do I Choose a VPN?

Start with *why*: what do you need a VPN for? VPNs serve different priorities:

» **Privacy-focused.** Choose one with a strict no-logs policy.

» **Streaming or bypassing geo-blocks.** If you need to change your browsing location, look for VPNs that explicitly support Netflix, Hulu, and so on.

» **Torrents/P2P.** Ensure the VPN allows torrents and has a kill switch.

» **Speed-sensitive tasks.** For gaming and video calls, you'll want one with fast servers and low latency.

Once you've addressed why you need a VPN, you can review the considerations below and choose the one that best suits your needs.

WHY THIS MATTERS

» **Data security.** A VPN provider with weak encryption or questionable logging practices can leave you worse off than not having a VPN, depending on your requirements.

» **User experience.** Clunky software or frequent disconnects lead many to abandon VPN usage.

» **Cost vs.** benefit. Some free VPNs harvest or sell your data, negating the point of using a VPN in the first place.

KEY CONSIDERATIONS

» **Reputation and logging policy.** Look for a "no-logs" policy, ideally audited by a third party.

- » **Server locations.** Having more servers in multiple locations typically means better speed and flexibility. But a service with servers in China, Russia, or North Korea may not provide the security or privacy you are seeking.
- » **Encryption and protocols.** Watch for modern protocols, such as WireGuard or OpenVPN.
- » **Multi-device support.** If you want to protect your phone, laptop, and tablet, ensure they can all run the VPN simultaneously.
- » **Due diligence.** It is your responsibility to verify that specifications align with your requirements.

IMMEDIATE ACTIONS

- » **Read reviews.** Check reliable tech outlets or user communities for honest feedback.
- » **Compare features.** Some VPNs focus on streaming, while others focus on raw privacy or speed. Choose what aligns with your goals.
- » **Test it.** Many VPNs offer trial versions or money-back guarantees. Try it for a week or two before making a commitment.

PUTTING IT ALL TOGETHER

Selecting the right VPN involves striking a balance between cost, privacy assurances, speed, and device compatibility. Aim for well-respected providers with transparent, third-party-audited no-logs policies to ensure your safety.

The Risk of Inbound VPN Connections on Your Network

Inbound VPN connections can expose your network to external systems, potentially allowing attackers to exploit vulnerabilities or compromised credentials to gain unauthorized access. They can bypass your security controls, increasing the risk of lateral movement and data breaches. Proper configuration, strict access controls, and robust monitoring are essential to mitigate these risks.

WHY THIS MATTERS

- » **Home servers and port forwarding.** Some advanced users set up inbound VPNs to connect remotely to their home network. This can be secure but also dangerous if misconfigured.

» **Attack surface.** Exposing a VPN server on your home router or PC effectively opens a door that hackers may try to exploit.

» **Access control.** Once someone is "inside" your network, it's often easier for them to move laterally and compromise other devices.

KEY CONSIDERATIONS

» **Strong configuration.** If you run an inbound VPN, it must be configured with robust encryption, strong credentials, and ideally multi-factor authentication.

» **Separate networks.** Placing your VPN server on a segmented portion (VLAN) of your network can limit the damage if it is compromised. You may not be able to configure this if it is part of your router configuration.

» **Updates and monitoring.** Regularly patch VPN server software and monitor logs for suspicious activity.

IMMEDIATE ACTIONS

» **Research thoroughly.** Follow best practices and official documentation to set up an inbound VPN for your specific device, such as a remote desktop.

» **Disable by default.** Don't enable inbound VPN unless you need remote access to your home devices.

» **Review router security.** Check if your router has built-in VPN features. If so, confirm they're secured with strong passwords and encryption protocols.

PUTTING IT ALL TOGETHER

Inbound VPN can be a powerful remote administration tool but is also a two-way street. Keep the door locked with modern encryption, strong credentials, and a vigilant eye on network logs. The need to connect to your home network may not be necessary if you use cloud-based file storage and tools. For example, storing files in OneDrive or Google Drive allows you to remotely access your files from anywhere in the world without the need for a VPN. Make sure you need it before you get it.

Which Devices Should Have or Use a VPN?

The following chart provides recommendations for the types of devices that can benefit from or need a VPN.

Device	VPN Recommended?	Why
PC/Laptop	Yes	High data exposure; full-featured VPN apps.
Phone/Tablet	Yes	Privacy, travel, mobile networks.
Smart TV/Streamer	Maybe	Only if geo-spoofing is needed.
Game Console	Rarely	Limited value unless masking IP for gaming.
IoT Devices	Only via router	No native support, not necessary.
Router	Yes	Full-network coverage, default privacy for all.

WHY THIS MATTERS

>> **Consistency.** If a VPN protects your laptop but not your phone, you could still be vulnerable when you switch devices.

>> **Public Wi-Fi use.** If permitted, smartphones and tablets can automatically connect to public Wi-Fi hotspots. A VPN is invaluable here.

>> **Resource usage.** Some older devices might struggle with the overhead required for encryption.

KEY CONSIDERATIONS

>> **Prioritize portable devices.** Laptops, tablets, and phones are at higher risk outside your home network.

>> **Smart TVs and streaming boxes.** Some people use VPNs to access geo-restricted content, but these devices may not have native VPN support, and your streaming service may not permit it.

>> **Desktop PCs.** A desktop computer may not require a VPN 24/7 unless you want to mask your IP address at home as well.

IMMEDIATE ACTIONS

>> **Start with mobile.** Install a VPN client on your phone and tablet if you frequently use public Wi-Fi, including guest wireless access at work.

>> **Check router-level VPN.** Some routers allow you to enable a VPN for the entire home network. This can protect all devices at once.

>> **Test performance.** Ensure your device can handle the VPN without significant slowdowns. If performance is an issue, consider upgrading your hardware or selecting a lighter protocol, such as WireGuard. Use speed test websites to check for performance hits using a VPN versus not using one.

PUTTING IT ALL TOGETHER

In an ideal scenario, all your devices would run behind a VPN whenever you're on unknown or public networks. However, if resources or complexities are a concern, at least cover your most at-risk devices, such as phones and laptops, when connecting outside your home's trusted environment.

Network Segmentation (VLANs)

Locking down each device on your network is critical, but it's only half the battle. By organizing your home or office network into separate "lanes," you can prevent unnecessary device interaction. This practice is known as **network segmentation**, which is often achieved through Virtual Local Area Networks (VLANs). While the term might sound a bit corporate, VLANs can be just as relevant and beneficial for home users who want an extra layer of security and organization for their growing collection of smart devices.

I'll explore VLANs, their differences from and relationships to VPNs, and when they are applicable. I'll also examine typical segmentation strategies, such as separating IoT devices from laptops or home office equipment, to reduce overall risk and maintain a clean network.

Picture an office building with multiple departments: finance, HR, and IT. You wouldn't want every HR employee to have direct access to sensitive financial records. The same logic applies to your network. When everything is lumped together, computers, smart TVs, and security cameras, an intruder who compromises one device may gain easy access to the rest. VLANs function like virtual walls, limiting the data and devices that are accessible to each other.

What's the Difference Between a VPN and a VLAN?

Understanding VLANs: Virtual Local Area Networks

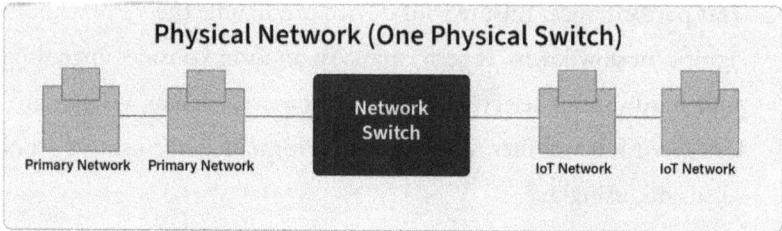

Physical Network (One Physical Switch)

Network Switch

Primary Network   Primary Network   IoT Network   IoT Network

With VLANs: Logical Network Separation

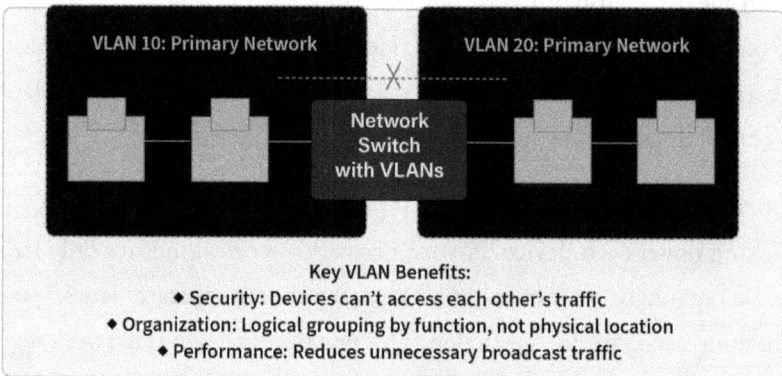

VLAN 10: Primary Network   VLAN 20: Primary Network

Network Switch with VLANs

Key VLAN Benefits:
◆ Security: Devices can't access each other's traffic
◆ Organization: Logical grouping by function, not physical location
◆ Performance: Reduces unnecessary broadcast traffic

Virtual Private Networks (VPNs) securely connect devices or networks over the internet, encrypting data to ensure privacy and security during transmission outside your home or office. Virtual Local Area Networks (VLANs) logically separate devices within a local network, like your home network, into distinct segments, managing internal traffic efficiently and improving security by isolating groups. Think of it this way: a VPN protects data traveling externally, while a VLAN organizes and isolates data internally.

WHY THIS MATTERS

» **Term confusion.** Many people mix up VPN and VLAN because they sound similar, but they serve different purposes.

» **Distinct functions.** VLANs segment your local (internal) network into separate zones, while VPNs create secure tunnels over the internet or external connections.

KEY DIFFERENCES

VPN (Virtual Private Network)	VLAN (Virtual Local Area Network)
Creates an **encrypted tunnel** for data traveling over the internet or other public networks.	Creates **logical partitions** within your local network, grouping devices or isolating them.
Often used for **secure remote access** or bypassing gee-restrictions.	Used to **segment** or **isolate** devices (e.g., guests vs. primary network, IoT devices vs. computers).
Typically **configured on endpoints** (e.g., laptops, phones) or at the router for establishing inbound and outbound tunnels.	Configured on a **managed switch** or **router** that supports VLAN tagging.
Focuses on **data confidentiality** across untrusted networks.	Focuses on **internal network organization and security**, specifically on which devices can communicate with each other.

IMMEDIATE ACTIONS

» **Clarify your goal.** Are you looking to protect data in transit over the internet (VPN) or to organize and limit traffic within your local area network (LAN) (VLAN)?

» **Check device support.** If you use VLANs, ensure your router or switch can handle VLAN tagging.

» **Set realistic expectations.** A VLAN won't encrypt your traffic as it exits the network, and a VPN won't isolate devices on your local area network (LAN).

Do VLANs and VPNs Work Together?

VLANs segment a local network into isolated groups for better management and security, while VPNs securely connect these segments (or remote users) across the public internet. Together, they provide controlled internal separation with secure external access. It's like organizing your home into separate rooms (VLANs) and securely locking doors when guests come over (VPNs).

WHY THIS MATTERS

» **Complementary tools.** VLANs and VPNs aren't competing technologies. They can function side by side to tackle different aspects of network security.

» **Advanced scenarios.** A small business might use VLANs to isolate guest Wi-Fi from internal systems and a VPN to secure remote access.

KEY CONSIDERATIONS

» **Layered approach (still here).** You can have multiple VLANs at home ("Main," "Guest," "IoT") and still tunnel out to a VPN when you're browsing. The VLAN ensures IoT cameras can't see your laptop's data; the VPN hides your external traffic from prying eyes.

» **Gateway configuration.** If your router supports both VLANs and a built-in VPN client or server, you can set rules that apply differently to each VLAN.

» **Remote access plus segmentation.** Some users configure inbound VPN access to a specific VLAN (such as a "work-from-home" VLAN) to keep it separate from their devices.

IMMEDIATE ACTIONS

» **Plan your topology.** If you have advanced needs, such as remote working, map out which VLANs require VPN access.

» **Set Access Control Lists (ACLs).** If your router supports it, define which VLANs can reach each other or the VPN tunnel.

» **Test thoroughly.** Complex setups can get confusing. Check that your segmentation rules and VPN settings don't inadvertently block essential traffic. Document your settings in case you have issues later.

Do I Need Both or Either?

The average home user doesn't need VLANs unless they're running multiple distinct networks — for example, IoT devices, one for children, another for work — or they have advanced home setups. A VPN, however, can benefit most users by protecting their privacy and data online. Basically, VLANs are usually not critical, but VPNs often are.

WHY THIS MATTERS

> » **Resource investment.** Setting up VLANs or VPNs can require new hardware or more advanced configuration.
>
> » **Right tool, right job.** Overcomplicating your network can cause headaches without tangible benefits.

KEY FACTORS

> » **Personal vs.** professional. A single-person household with minimal devices may not require VLAN segmentation, but a large home or small office with numerous IoT devices might.
>
> » **Risk appetite.** If you're handling sensitive work files from home, consider combining VLANs (to isolate your personal or entertainment devices) with a VPN (to connect to your workplace securely).
>
> » **Technical comfort.** VLAN setups can be intimidating if you're unfamiliar with networking. VPNs can be easier to enable, especially if offered as a simple router or software option.

IMMEDIATE ACTIONS

> » **Evaluate your devices.** If you have multiple IoT devices (such as a smart fridge, cameras, and doorbells), VLAN segmentation can contain them in case one is compromised.
>
> » **Consider your work requirements.** A VPN may be necessary if remote access is crucial, especially if your employer mandates it.
>
> » **Consider complexity.** If you aren't tech-savvy, it might be wise to keep it simple. VLANs or advanced routing could require specialized knowledge or professional help.

Why You Would Create a VLAN

You'd create VLANs at home to separate less secure devices — such as smart-home gadgets — from your main computers to reduce security risks or to isolate guest devices so visitors don't access your files. They can also prioritize traffic for streaming or gaming, enhancing network performance.

The basic idea is to partition your network into different "segments" that can't all freely talk to each other. Here are some common scenarios:

GUEST NETWORK

» **Why?** Let visitors use your internet without seeing your files or printers.

» **Devices.** Phones or laptops belonging to guests are separated from your primary network's shared drives.

IOT DEVICES

» **Why?** Many IoT gadgets have questionable security. You don't want it snooping on your main PC or phone traffic if one gets hacked.

» **Devices.** Smart TVs, doorbells, thermostats, cameras, appliances, and other devices that communicate with the cloud but do not require direct access to your workstation.

HOME OFFICE

» **Why?** Keep work data separate from the family's entertainment devices.

» **Devices.** Work laptop, network-attached storage (NAS) for professional files, VoIP phone. This can live on a VLAN restricted to authorized users only.

» Kids' VLAN

» **Why?** If you want to implement parental controls or limit online gaming traffic to prevent it from degrading the entire network, a kids' VLAN can help.

» **Devices.** Children's tablets, consoles, or laptops.

HOW TO IMPLEMENT

» **Managed router or switch.** You typically need hardware that supports VLAN tagging (802.1Q). Many "prosumer" routers now feature built-in VLAN options.

» **Naming.** Name or label each VLAN ("MAIN," "GUEST,"

"IOT," "WORK").

» **Firewall rules.** Determine whether VLANs can communicate with each other or only with the internet. For instance, the IoT VLAN might only have outbound internet access, which is entirely blocked from your main VLAN.

WHY THIS MATTERS

» **Containment.** If one VLAN is compromised, attackers can't pivot as quickly to your more sensitive devices.

» **Performance.** Separating bandwidth-intensive devices can help optimize traffic, especially if you can prioritize specific VLANs using Quality of Service (QoS). QoS is beyond the scope of this book.

» **Simplicity vs.** security. VLANs add complexity but significantly boost overall protection.

IMMEDIATE ACTIONS

» **Inventory devices.** Note which gadgets handle critical data (PCs, servers) and which are less trustworthy (IoT).

» **Research router capabilities.** Verify if your existing router or switch supports VLAN implementation. If not, you might need an upgrade.

» **Start small.** You don't need 10 VLANs. Even creating 1 or 2 ("MAIN" and "GUEST/IoT") can make a big difference.

PUTTING IT ALL TOGETHER

Securing your home network doesn't require advanced technical skills. You enhance your defense by changing default admin credentials, enabling strong encryption, disabling remote management, setting up a guest network, and monitoring connected devices. These steps form a crucial foundation that supports all other security measures, from protecting IoT devices to implementing safe smartphone practices.

Once these measures are in place, you'll find your home network far less vulnerable to common threats. You'll also be well-prepared to expand your cybersecurity efforts, whether that involves setting up a more advanced firewall or considering a Virtual Private Network (VPN).

Think of your secured router as your home's digital perimeter guardian. When properly configured, attackers often find it easier to move on to a less-protected target than attempting to penetrate your robust defenses. Following the straight-forward steps in this chapter, you're signaling to potential intruders: "This home network is locked down — don't bother."

With your network fortified, you're ready for the following chapters, which delve into protecting the growing array of smart devices populating our homes and managing the keys to your digital kingdom: passwords and account credentials. It's all part of a holistic (layered) security approach designed to keep your digital life — or, as I call it, your digital castle — standing strong against the ever-present waves of online threats.

A VPN can be a game-changer for online privacy and security, creating an encrypted tunnel that shields your data from prying eyes. You may or may not need one 24/7, but a VPN is essential if you regularly travel or use public Wi-Fi. Choosing the best service involves balancing cost, speed, reliability, and privacy policies and ensuring it works across all your devices.

Remember that inbound VPN connections to your home network demand extra caution. While remote access is convenient, misconfiguration can open a back door for malicious actors. Set up strong encryption and solid credentials and monitor any remote connection attempts.

Additionally, consider your entire device ecosystem. You're still exposed if you protect your laptop but leave your phone or tablet unencrypted. With thoughtful selection, proper setup, and vigilance, VPNs can keep your online activities under wraps and make those coffee-shop Wi-Fi sessions much less nerve-wracking.

Network segmentation with VLANs is like erecting internal walls in your digital home. Instead of allowing every device to see every other device, you group them by function or trust level, thereby minimizing lateral movement if a single device is compromised. Although commonly associated with business environments, VLANs offer valuable security and organizational benefits in a home network, especially for those with various IoT devices.

Also keep in mind that VLANs solve a different problem than VPNs. A VPN protects data in transit over untrusted networks, such as the internet, while VLANs focus on local network isolation. They can coexist in a layered defense strategy, such as multiple VLANs within your home, along with a VPN for secure remote access or online anonymity.

You must determine whether you have sufficient devices or sensitive data to justify VLAN segmentation. Do you need remote access or to mask your external internet traffic, which would require a VPN? Your specific setup and security needs will dictate whether you implement neither, one, or both. If done correctly, VLANs can transform your network from an open-plan room into a smartly partitioned space where each device stays in its designated lane. ⁑

CHAPTER FOUR

Authentication

AT ITS CORE, AUTHENTICATION IS THE process of proving who you are. For cybercriminals, every login attempt is an opportunity to infiltrate someone else's account, steal data, or impersonate them online. Strong authentication processes dramatically reduce your risk of being compromised, whether you're signing into social media or an enterprise database.

In this chapter, I'll examine the primary authentication methods: passwords, passkeys, hardware keys, biometrics, and authenticator apps, exploring how they function, their various types, and their corresponding security levels. I'll then consider the pros and cons of each, show you how to layer them for more vigorous defense, and explain how to determine if your credentials have been compromised in a breach.

In short, it's your guide to securing the front door of your digital house more effectively and adding a few extra layers of protection to keep intruders out.

The Case for Strong Authentication

Picture someone casually picking the lock on your front door because the lock is flimsy or because you left the key under the doormat. In the digital realm, a weak or single-factor login is an insecure door. Conversely, layering *(I'm still here)* multiple factors, such as a sturdy deadbolt and an alarm system, forces attackers to jump through extra hoops.

Strong authentication is more than just picking a complex password. It involves leveraging different factors: what you know, what you have, or what you are. Even if one fails, the others stand firm. Let's explore the primary options available to you.

The Different Types of Authentication Methods

Why multiple methods? Each has strengths and weaknesses. Combining them improves overall security by reducing reliance on a single vulnerable method. Below are five primary methods of proving your identity online: passwords, passkeys, hardware keys, biometrics, and authenticator apps. I'll break down how each works, the various forms they can take, and how secure they are in practical use.

WHY THIS MATTERS

» **Multiple options.** You're not stuck with just "username and password." Innovations, such as hardware tokens, biometrics, and authenticator apps, significantly reduce reliance on memorized secrets.

» **Context-specific.** What works for your email might differ from what's best for critical work systems.

» **Varying levels of security.** Each method has its strengths and weaknesses. Knowing these lets you tailor a solution to each account's importance.

Method One: Passwords

» The classic method; something you know.

» Usually the easiest to implement, but also one of the most attacked if not well-managed.

HOW THEY WORK

» **Definition.** A password is a secret string of characters (letters, numbers, symbols) that only the legitimate user is supposed to know.

» **Authentication factor.** Something you know.

TYPES

» **Simple passwords.** Basic, often short ("pass123"); easily guessed or brute forced.

» **Complex passphrases.** Uses longer, more intricate strings or phrases ("BlueMoon#37WentWest!"); much harder to crack.

» **Managed passwords.** Created and stored by a password manager that can generate unique, long, random passwords for each account.

SECURITY LEVEL

- » **Highly variable.** A short or reused password is extremely risky; a long, random passphrase stored in a manager is significantly safer.
- » **Vulnerabilities.** It is susceptible to brute force, phishing attacks, and guesswork if poorly chosen.

Method Two: Passkeys (phone or browser-based)

- » **A feature on your device, often used in conjunction with a PIN or biometric.**
- » **Eliminates the need to memorize complex strings and can be more resistant to phishing.**
- » **If the device is lost or stolen, you'll need a reliable backup method.**

HOW THEY WORK

- » **Definition.** Passkeys are cryptographic login credentials often tied to your device (such as a phone or laptop) and protected by a local PIN or biometric measure.
- » **Authentication factor.** Something you have (the device), possibly something you are (biometric measure), or something you know (PIN) to unlock it.

TYPES

- » OS-integrated passkeys.
- » Built into platforms such as iOS, Android, or specific browsers.
- » Cross-platform passkeys.
- » Synced across multiple devices via cloud services. For example, you can set up a passkey on your phone and use it on your tablet as well.

SECURITY LEVEL

- » **Generally high.** Resistant to phishing because each service gets a unique credential.
- » **Risk.** If someone steals or compromises your device, they may attempt to bypass your device's local PIN or biometric lock. Always maintain a strong device-level lock.

Method Three: Hardware Security Keys

» A physical device you have, such as YubiKey.

» Typically plugged into a USB port (or connected via NFC) when logging in.

» Very difficult to compromise remotely but can be misplaced or lost.

HOW THEY WORK

» **Definition.** A small physical token (often USB or NFC-based) that communicates with a site or service to confirm your identity.

» **Authentication factor.** Something you have; the key itself.

TYPES

» USB-A, USB-C Keys.

» Plug into a computer port for desktop authentication.

» NFC/Bluetooth keys.

» Tap against or pair with a smartphone or laptop that supports NFC or Bluetooth for wireless sign-in.

SECURITY LEVEL

» **Very high.** An attacker must possess the physical key to break in, making remote hacking extremely difficult.

» **Weaknesses.** Physical loss or damage, plus not all services support hardware keys. You need a solid backup plan in case you lose it.

Method Four: Biometrics

» **Something you are.** For example, your thumbprint, face ID, or iris scan.

» **Easy to use.** Once it's set up, there is no password to remember.

» **Privacy concerns.** Storing biometric data is not foolproof, although it's still challenging to fake.

HOW THEY WORK

» **Definition.** Relies on unique biological traits, such as fingerprints, facial recognition, and iris scans.

» **Authentication Factor.** Something you are.

TYPES

» **Fingerprint scanners.** Common on phones and laptops and scans patterns on your fingertip.

» **Facial recognition.** Utilizes your device's camera and infrared sensors to create a 3D map of your face.

» **Iris or retina scans.** It is primarily used in high-security settings, such as enterprises, and scans eye patterns.

SECURITY LEVEL

» **Generally high.** Very convenient and relatively challenging to fake.

» **Potential flaws.** Sophisticated spoofing techniques can trick facial recognition systems, and concerns exist about the storage and security of biometric data.

Method Five: Authenticator Apps

» An app typically installed on your smartphone or another device that generates time-based one-time passcodes (TOTPs). Google Authenticator, Microsoft Authenticator, and Authy are popular authenticator apps.

» The app generates a new six- or eight-digit code every 30 seconds. When you log in, you enter your password plus this one-time code.

» Safer than relying on text messages (SMS), as SIM swaps and phone carrier hacks are common attack vectors.

HOW THEY WORK

» **Definition.** A smartphone or device app that generates one-time passcodes (often referred to as TOTPs, or Time-Based One-Time Passwords) for each login.

» **Authentication factor.** Something you have (such as your phone or device).

TYPES

Vendor-specific

» For example, Google Authenticator, Microsoft Authenticator, or Adobe.

» Integrate seamlessly with their respective ecosystems and possibly others.

Independent

» For example, Authy or Duo Mobile.

» Offers multi-device sync and backups across platforms and can manage multiple accounts.

» Some authenticator apps are restricted to specific platforms. Ensure the one you choose is compatible with various platforms and operating systems, as well as your devices.

Password manager-integrated

» Some password managers also generate TOTPs, consolidating all authentication information in one place.

SECURITY LEVEL

» **High.** Codes refresh every 30 to 60 seconds, making them difficult to intercept. They are not tied to your phone number, so they are immune to SIM-swap scams.

» **Risks.** You could be locked out if you lose your phone without backup codes. Always keep a fallback method.

PROS AND CONS OF EACH METHOD

Method	Pros	Cons
Passwords	Familiar. Widely supported. Easy to update or change.	Often weak if resued. Vulnerable to phishing and brute force attacks. Easy to forget.
Passkeys	Eliminates remembering long passwords. More phishing-resistant. Convenient oif the device is always with you.	Device dependent; risk if stolen. Setup and syncing can be tricky. Not universally supported.

Method	Pros	Cons
Hardware Keys	Extremely secure. Highly resistant to remote attacks. Great for high-risk stakes accounts.	Susceptible to loss or damage. Additional cost. Not all services support them.
Biometrics	Fast and convenient. No memorization needed. Common in modern phones, laptops, and tablets.	Privacy concerns. Some methods can be spoofed. If compromised, you can't "change" your fingerprint.
Authenticator Apps	Strong against phishing and SIM-swaps. Broad service support. Codes expire quickly.	Requires careful backup and/ or transfer. Dependent on your device. If lost or broken, you have big trouble.

Layering Authentication for Added Security

Multi-factor authentication (MFA) combines at least two factors — something you know, have, or are — to form a robust defense. If a hacker steals your password, they'd still need your phone for the authenticator app or your physical hardware key, making the breach far less likely.

» **Pair a strong password with an authenticator app.** Common, user-friendly. Even if your password is leaked, the one-time code prevents unauthorized logins.

» **Use biometrics plus a hardware key.** Ideal for crucial systems or corporate logins. The intruder would need your fingerprint or face, along with your physical token.

» **Enable passkeys and backup codes.** If your device supports passkeys, store backup codes separately in case the device is stolen or reset.

Have You Been Breached?

Checking if you're part of a data breach is crucial because compromised information often includes sensitive personal data, passwords, or financial details that criminals exploit for identity theft, financial fraud, or account takeover. Ignoring breach notifications significantly increases your vulnerability to cyberattacks, potentially leading to financial losses, damaged credit, and serious privacy violations.

WHY THIS MATTERS

» **Immediate risk.** A leaked password or phone number can become an open door for attackers.

» **Proactive defense.** If you discover a breach quickly, you can change passwords or enable additional factors before bad actors exploit the leak.

» **Password reuse.** Using the same password across multiple sites increases the risk of a single breach affecting all your accounts and devices.

HOW TO CHECK

» **Online breach databases.** Sites like haveibeenpwned.com allow you to enter your email or phone number to check if it appears in known breach databases.

» **Dark web monitoring.** Some security suites or identity protection services scan underground marketplaces.

» **Alerts and news.** Stay tuned for official notices from the services you use. If a breach occurs, they'll often prompt users to reset passwords.

IMMEDIATE ACTIONS

» **Reset passwords.** Begin with the ones on the breached site and then proceed to any accounts with similar credentials. *Reminder: use unique credentials for every site or application.*

» **Enable or update MFA.** Add an authenticator app, hardware key, or passkey if you haven't already done so.

» **Check logins.** Look for suspicious activity, such as login attempts from unknown locations.

Managing Passwords and Account Security

Your passwords are the keys to your digital life, unlocking access to everything from your email and social media accounts to your bank accounts and online shopping platforms. With so much valuable information tied to these credentials, securing your passwords is essential for protecting your privacy, finances, and personal data. Fortunately, while strong password management may seem complex, it is one of the most straightforward ways to significantly enhance your security.

Here, I'll cover best practices for managing passwords effectively, ensuring your online accounts are as secure as possible. From creating unique passwords for every account to utilizing password managers and enabling two-factor authentication (2FA) or multi-factor authentication (MFA), these steps will equip you with the tools you need to safeguard your digital identity.

The Importance of Password Security

In the modern world, your password is often the only line of defense between your personal information and cybercriminals. If someone gains access to your passwords, they can wreak havoc by stealing your financial details, impersonating you online, or accessing sensitive work or personal data. Cybercriminals employ methods such as phishing attacks, keylogging, and brute-force attacks to crack weak or reused passwords. Adopting strategies that strengthen and protect your passwords is crucial to keeping you one step ahead of potential attackers.

Use Unique Passwords for Every Account

One of the more common and dangerous mistakes people make is reusing the same password across multiple accounts. While this might seem convenient, it increases your vulnerability exponentially. If one of your accounts is breached, the hacker can access all of your accounts that share the same password.

WHY THIS MATTERS

If you use the same password for multiple accounts, all it takes is one breach for a hacker to gain access to several services at once. For example, if your email password is compromised, hackers could access your online banking, shopping sites, and even social media accounts, potentially using them to steal your money or reputation.

By ensuring each account has a unique password, you drastically reduce the impact of a potential data breach on any account.

STEP-BY-STEP GUIDE

- » **Create unique passwords.** For each account, generate a password specific to that account and service. Avoid using common passwords such as "123456" or "password123."

» **Don't reuse passwords.** Never reuse the same password for multiple accounts, even if you're tempted by convenience. Each password should be distinct and not linked to another service.

» **Document your passwords securely.** Write each password down in a safe place (more on this below) or store them securely in a password manager.

BEST PRACTICES

» **Password length over complexity.** Long passwords that mix letters, numbers, and symbols are more secure than short, complex ones.

» **Use phrases.** Instead of a single word, try using phrases like "Love$Dogs_123!". Longer phrases are more challenging to crack than individual words.

Use a Password Manager

A password manager helps you generate, store, and auto-fill complex passwords for your online accounts and locally installed applications. It's a simple way to keep track of all your passwords without having to remember each one manually. Using a password manager makes your life easier and significantly enhances your security by ensuring your passwords are both strong and unique.

WHY THIS MATTERS

Creating unique passwords for every account is essential, but remembering all those passwords can be difficult, if not impossible. A password manager securely stores your passwords in one place, encrypting them for added protection. It also generates random, strong passwords, eliminating the need to guess or create your own.

STEP-BY-STEP GUIDE

» **Choose a password manager.** Some popular options include LastPass, 1Password, RoboForm, and Bitwarden. Many have both free and paid versions, but the free versions typically come with restrictions that limit their functionality. You will need to exercise due diligence for this part. Many choices exist, and the feature sets and compatibility differ greatly. Be sure to compare features and choose the one that most closely matches your needs and is reputable. Choose a password manager that supports syncing across devices, allowing you to access your passwords from your phone, tablet, or computer.

» **Set a strong master password.** Your master password is the key to your password vault. Make it long, complex, and something you won't forget (for example, use a combination of random words and characters). While you may be able to store the master password in the password manager, if you forget it, everything is lost. You mustn't lose or forget your master password.

» **Store and generate passwords.** The password manager will store all your passwords securely and allow you to generate new, strong passwords for every new account.

» **Enable Two-Factor/Multi-Factor Authentication (2FA/MFA) on the password manager.** Ensure that your password manager itself is protected with 2FA/MFA. This adds an extra layer of security (I warned you) if someone gains access to your master password.

BEST PRACTICES

» **Use strong encryption.** Always choose a password manager that encrypts your data, so even if someone gets access to your vault, they can't view your passwords.

» **Don't forget your master password.** The usefulness of your password manager depends on your ability to remember this one key password. Write it down securely and store it offline if needed.

Enable 2FA or MFA

Two-Factor Authentication (2FA) and Multi-Factor Authentication (MFA) add an extra layer of security to your accounts by requiring something beyond just your password. With 2FA, even if someone gains access to your password, they won't be able to log in to your account without also possessing the second factor. MFA means any security protocol that involves two or more authentication steps. I've simplified these terms to MFA/2FA for this discussion.

MFA/2FA is one of the easiest and most effective ways to add another layer (again?) of security to your accounts, making it much harder for hackers to gain unauthorized access.

WHY THIS MATTERS

While passwords are a crucial security measure, they can still be compromised through data breaches, phishing, or brute-force attacks. MFA/2FA provides an additional layer of verification, typically something only you can access, such as a smartphone or an authentication app. The authentication keys are either something you uniquely know or have.

STEP-BY-STEP GUIDE

» **Go to account settings.** Open the account settings for the service or website you want to protect, such as your email, social media, or banking accounts. Look for the "Security" or "Login" section.

» **Enable MFA/2FA.** Look for the option to enable MFA/2FA. This is often located in a subsection called "Security Settings" or "Account Settings." Follow the prompts to set it up, including receiving a text message with a code or linking an authentication app such as Google Authenticator or Microsoft Authenticator.

» **Choose your second factor.** The second factor can be a text message (SMS), an authentication app (for better security), or even a hardware security key.

» **Create backup codes.** Some services offer backup codes if you lose access to your 2FA method. Write your codes down and store them safely, such as in a password manager.

BEST PRACTICES

» **Use authenticator apps.** Whenever possible, use an authenticator app instead of SMS-based MFA/2FA. Text messages can be intercepted or hijacked.

» **Enable MFA/2FA on all accounts.** Prioritize enabling MFA/2FA for your most sensitive accounts, such as email, banking, and social media. I recommend using MFA/2FA wherever it is offered.

Watch Out for Security Questions

Many online services use security questions as an additional layer of identity verification. While these are designed to help you recover accounts, attackers can often exploit them if you're not careful with your answers.

WHY THIS MATTERS

Security questions, such as "What is your mother's maiden name?" or "What was your first pet's name?" are often easily guessable or searchable on social media. If attackers can determine the answers to these questions, they may be able to reset your password and gain access to your account.

STEP-BY-STEP GUIDE

» **Choose strong, non-obvious answers.** For security questions, avoid using answers that could be easily found on your social media profiles or that are publicly available. Instead of answering truthfully, consider using random answers, which you can store securely in your password manager.

» **Use a password manager for security questions.** Many password managers also allow you to store answers to security questions. Treat these answers like passwords and make them random and long.

» **Avoid common questions.** If the service allows, choose security questions that are less obvious or have unique answers only you would know.

BEST PRACTICES

» **Randomize answers.** Make your answers as obscure as possible. For example, answer "What is your favorite color?" with "Purple@39!"

» **Use password managers.** Your password manager can help generate and store random answers to security questions, making it easier to stay secure.

While security questions are designed to verify your identity, they can also pose a significant risk if the answers are easily guessed. Protect them with random, secure responses.

PUTTING IT ALL TOGETHER

Authentication is the bedrock of cybersecurity: no matter how tight your anti-virus or encryption, if anyone can log in as you, the rest crumbles. By combining modern methods, such as passkeys, hardware keys, biometrics, and authenticator apps, each account can achieve the security level it deserves.

Remember the final puzzle piece: vigilance. Even the most robust authentication setup can be compromised if your credentials are leaked in a data breach and you fail to take action. When breaches occur, check for compromised accounts and rotate passwords. Reset your authenticator app, if needed. With a thoughtful, layered strategy — preferably one that leans on methods safer than text messages — your digital front door will remain locked against most online threats.

Managing your passwords and account security is one of the most important aspects of safeguarding your digital life. By using unique passwords for each account, utilizing a password manager, enabling two-factor or multi-factor authentication, and being cautious with security questions, you can build a strong defense against unauthorized access.

The key is consistency; once you establish these practices, they become second nature. With some initial setup, you can manage your accounts confidently, knowing you've implemented the best strategies to protect your personal information. By remaining vigilant and adopting new security measures as needed, you'll be well-prepared to face the evolving threats of the digital world. ⁑

Patching and Updates

MANY ATTACKS EXPLOIT OUTDATED SOFTWARE. PERFORMING regular updates is a simple but powerful defense. One of the most fundamental — yet often overlooked — defense mechanisms in cybersecurity is keeping software up to date. Whether it's your operating system, mobile apps, or firmware on IoT devices, regular updates are your first defense against potential security vulnerabilities.

Cybercriminals actively seek out outdated systems and unpatched software that are vulnerable to exploitation. The good news is that staying up to date is simple, effective, and critical for securing your devices. This chapter explains the importance of regular patches and updates, providing clear, actionable steps to ensure your systems are always protected.

Why Patch and Update?

Software vulnerabilities are often discovered after a product is released. Hackers can exploit these vulnerabilities to gain access to your systems, steal data, or cause damage. When software companies become aware of these vulnerabilities, they release patches and updates to fix the issue. However, if you don't install these updates, your devices and systems remain exposed to attacks. Regularly installing patches and updates significantly reduces the risk of falling victim to cyberattacks.

» **Security patches.** Updates typically contain fixes for known security holes. Neglecting updates leaves those vulnerabilities open for exploitation.

» **Performance improvements.** Many updates also enhance performance, speed, and usability, ensuring that your devices continue running smoothly.

> **New features.** Some updates introduce valuable new features or tools that enhance the functionality of your device.

Turn On Automatic Updates

Automatic updates are the easiest way to ensure your devices are always protected. By enabling automatic updates, you eliminate the need to manually track when updates are available. This saves you time and ensures that your systems stay up to date without you needing to remember to check.

WHY THIS MATTERS

Many cyberattacks exploit vulnerabilities in outdated software. Automatic updates ensure that patches are installed as soon as they become available, reducing the window of opportunity for hackers to exploit vulnerabilities in your system.

STEP-BY-STEP GUIDE

Operating systems

> **Windows.** Go to Settings > Windows Update and turn on "Get the latest updates as soon as they're available.".
> **Mac.** Go to System Settings > General > Software Update > Automatic Updates and move the four sliders in the Automatically pop-up section to the right. Click Done.
> **iPadOS.** Go to Settings > Software Update > Automatic Updates and toggle the three sliders to the right to enable.
> **Linux.** On most Linux distributions, updates are automatically handled through package managers, but you can configure automatic updates through system settings.

Apps and software

> **Android.** Go to the Google Play Store > Click on your account icon > Settings > Network Preferences > Auto-update apps and select your preferred method for downloading updates.
> **iPadOS.** Go to Settings > Apps > App Store and move the App Updates slider to the right.
> **Desktop software.** For software like browsers, office suites, and other

productivity tools, ensure that auto-update options are enabled.

Firmware updates for devices

» **Smart devices.** For IoT devices, routers, and other connected devices, check the settings or app associated with the device to ensure that firmware updates are enabled.

BEST PRACTICES

» **Set a schedule.** While automatic updates ensure timely installation, you can also schedule them to run during off-hours to avoid interruptions.

» **Don't ignore restart prompts.** Some updates require a system restart. Ensure that you complete these to finalize the installation.

Enabling automatic updates ensures that your devices receive essential security patches as soon as they become available, protecting you from known vulnerabilities.

Manually Check for Updates

While automatic updates are convenient, not all devices and applications update automatically, especially more specialized hardware such as routers, printers, or specific IoT devices. In these cases, it's essential to manually check for updates to ensure your systems remain protected.

Manually checking for updates ensures that devices and software that don't update automatically are kept current, reducing your exposure to security risks.

WHY THIS MATTERS

If updates are ignored, devices that don't automatically update may be left vulnerable to threats. You can ensure you're not exposed to known security risks by proactively checking for updates.

STEP-BY-STEP GUIDE

Routers and modem firmware

» Log into your router's admin interface (typically via a web browser using the router's IP address). Check for the "Firmware Update" or "System Update" section. If available, download and install the latest version.

<u>Smart devices and appliances</u>

> » For devices like smart cameras, thermostats, or doorbells, check the manufacturer's app or settings to verify if firmware updates are available and how to install them.

<u>Other software</u>

> » For desktop apps, check the app's Help or Settings menu for an option to "Check for Updates." Many apps will notify you when updates are available, but it's a good practice to check periodically.

BEST PRACTICES

> » **Check monthly.** Set a reminder to manually check for updates for devices that don't automatically update, such as routers, smart devices, and specific software tools. You should also check your other devices and software to ensure that automatic updating is occurring.

> » **Prioritize security updates.** Always install security patches as soon as they are available, even if they seem minor. The latest updates often contain critical security fixes.

Replace Old, Unsupported Devices

Outdated hardware, such as older smartphones, routers, or printers, may no longer receive security updates from the manufacturer. Using unsupported devices can leave your network and data vulnerable to attacks. The same is true of software. Everything has a lifespan. Replacing outdated devices is essential to maintaining a secure environment. It ensures that your hardware can continue to receive critical updates and patches.

WHY THIS MATTERS

Once a manufacturer discontinues support for a device or software, it ceases to release security patches, leaving the device vulnerable to new threats. This is especially important for devices that store or transmit sensitive data.

STEP-BY-STEP GUIDE

> » **Identify outdated devices.** Check the manufacturer's website or support pages to see if your device is still supported. If the device

no longer receives updates or security patches, it's time to consider upgrading.

» **Upgrade to newer models.** For devices like routers, smart home products, or smartphones, look for newer models that are still supported by the manufacturer and receive regular updates.

» **Replace software, not just hardware.** If you're using outdated software or apps on supported hardware, ensure they're updated. Unsupported software can pose security risks even on a newer device.

BEST PRACTICES

» **Plan for upgrades.** Set aside a budget for upgrading essential devices, particularly those connected to your home or office network, such as routers and network storage devices.

» **Check compatibility.** Ensure that new devices are compatible with your existing systems and software to avoid disruptions during the upgrade process.

PUTTING IT ALL TOGETHER

Patching and updates are one of the simplest and most effective defenses against the constant threat of cyberattacks. By turning on automatic updates, manually checking for updates on devices that don't update automatically, and replacing outdated hardware, you ensure your devices remain secure and protected from known vulnerabilities.

Staying up to date on software patches and device updates is a continuous process, but it's one of the most powerful ways to defend against security breaches. Make updates and patches a routine part of your device maintenance, and you'll significantly reduce the likelihood of falling victim to attacks that exploit outdated software.

Regularly updating and upgrading your devices helps keep both your digital and physical environments secure, ensuring peace of mind and minimizing the risk of cyber threats. ⁑

CHAPTER SIX

Anti-Virus and Anti-Malware Tools

LIFE WOULD BE SIMPLER IF CYBER threats wore bright neon signs saying, "Hey, I'm dangerous, click at your own risk!" Instead, viruses and malware often infiltrate systems by disguising themselves as harmless links or attachments, waiting for unsuspecting users to allow them in unwittingly. This is why anti-virus (AV), and anti-malware tools remain crucial. they serve as your digital guard dogs, sniffing out threats that would otherwise be invisible.

In this chapter, I'll explore different types of security software, how to choose the right one, and the basic upkeep (like updates and scans) that keeps your defenses ready for the next big threat. I'll also touch on what happens when suspicious items get quarantined and how to decide whether flagged files are safe or malicious. This isn't just about installing software and hoping for the best; like every other layer of security, it needs a little attention and regular TLC (tech-loving care).

The Importance of Security Software

Although modern operating systems come with built-in defenses, malware continues to evolve. Criminals aren't taking breaks to admire your new laptop; they're constantly engineering new ways to infiltrate systems. Antivirus/anti-malware tools provide an active layer of protection, scanning files as you download or install them, monitoring suspicious behavior, and quarantining any harmful content. Think of them as digital bodyguards, always lurking in the background to intercept trouble before it escalates.

This chapter will guide you through the key categories of antivirus and antimalware solutions, as well as how to maintain them effectively. I'll also delve into how to interpret "quarantine" alerts, as not every flag is a death sentence for your files.

WHY THIS MATTERS

» **First line of defense.** Mistakes can still occur, even with safe browsing and email habits. A momentary lapse can invite malware in.

» **Automated protection.** Modern AV solutions update automatically, scanning for threats with minimal user intervention.

» **Peace of mind.** Knowing you have a digital watchdog can help lower stress and allow you to focus on everyday tasks without constant worry.

KEY CONSIDERATIONS

» **Comprehensive coverage.** Some malware can slip past essential built-in protection. Third-party or specialized software might offer more substantial protection and more frequent updates.

» **System compatibility.** Different operating systems (Windows, macOS, Android) have varying built-in defenses. Determine if you require a more robust solution.

» **User awareness.** Even the best tools can't protect against every possible threat. Combine them with safe browsing habits for maximum impact.

IMMEDIATE ACTIONS

» **Check what's already installed.** Many devices have a default security app. Make sure it's activated and up to date.

» **Research options.** Consider whether a more robust tool is needed. Free and paid solutions often differ in features such as real-time scanning, ransomware protection, and identity theft monitoring.

» **Educate yourself.** Understand basic terms like "virus," "spyware," "ransomware," and "trojan." Recognizing these helps you interpret alerts and news about threats.

PUTTING IT ALL TOGETHER

Security software is more than just a "nice-to-have." In a world where data is currency, ensuring you have at least one layer of protection is essential. Combine this software with thoughtful online behavior, such as not clicking on random email links or installing unknown apps to significantly reduce your risk of infection.

The Different Types of AV Products and the Pros and Cons of Each

Understanding different antivirus products and their pros and cons helps you choose the protection that best fits your security needs, device compatibility, and budget. Not all antivirus software offers equal protection; knowing its strengths and weaknesses ensures your devices stay secure without wasting resources on unnecessary features.

WHY THIS MATTERS

- » **Variety is overwhelming.** A quick online search reveals countless "Top 10 Anti-Virus" lists. Understanding each category helps you narrow down what's right for your needs.
- » **Feature differences.** Some solutions detect traditional viruses, while others focus on protecting against ransomware or phishing attacks.
- » **Budgeting.** Paid software often includes additional features such as VPNs, password managers, or parental controls. But if you only need basic scanning, a free solution might suffice.

KEY CONSIDERATIONS

- » **Traditional antivirus.** Focuses on known signatures, which is excellent for common threats but less effective against new malware.
- » **Next-Gen and AI-powered tools.** Utilize behavioral analysis to identify suspicious activity. These tools are helpful for zero-day or unknown threats, though they may generate more "false positives."
- » **Free vs. paid.** Free versions often have fewer features, such as real-time scanning or comprehensive support. Paid suites can provide all-in-one solutions but may feel bloated if you need the basics.

IMMEDIATE ACTIONS

- » **Assess your risk.** If you're a casual home user, a reliable free tool combined with built-in OS security might be sufficient. Power users or those handling sensitive data might need advanced solutions.
- » **Check reviews and benchmarks.** Independent testing labs, such as AV-TEST and AV-Comparatives, regularly publish performance and

detection rates.

» **Test drive.** Many paid products offer trial periods; try them out before you commit.

PUTTING IT ALL TOGETHER

There is no one-size-fits-all answer. The best AV/anti-malware tool depends on your usage habits, the type of data you handle, and your budget. A well-reviewed, free solution is often better than no protection at all, but if you need specialized features (such as anti-ransomware or advanced web filters), investing in a paid suite might be worthwhile.

How to Choose the Best Antivirus or Anti-Malware Software

Choosing the proper antivirus or anti-malware software is crucial because not all products provide the same level of protection. A strong solution safeguards your devices against evolving threats like viruses, ransomware, spyware, and phishing attacks, lowering the risk of identity theft, data loss, and financial harm. Skimping on protection is akin to buying a cheap lock for your front door: you might save a few dollars initially, but the actual cost is felt when someone breaks in.

WHY THIS MATTERS

» **Customization.** Not everyone has the exact device count, data sensitivity, or risk tolerance. A solution that works for a home-based freelance writer may differ from the one used by a traveling sales rep.

» **Avoiding overkill.** Overly complex tools can slow down your system, hinder productivity, or cause user frustration, leading people to disable them, which is a big mistake.

» **Long-term fit.** Switching solutions frequently can be a hassle. A careful choice now can save time and headaches later.

KEY CONSIDERATIONS

» **System performance.** Some antivirus solutions require more system resources. Older machines might struggle with robust scanning engines.

» **Ease of use.** If you're not tech-savvy, look for intuitive dashboards.

You don't want to wade through hidden menus to run a quick scan.

» **Support and community.** A good support forum or user community can be invaluable when facing a tricky infection or false alarm.

IMMEDIATE ACTIONS

» **Make a feature wish list.** Do you need parental controls, a firewall, or VPN integration? Separate your must-haves from your nice-to-haves."

» **Check compatibility.** Ensure the tool supports all your devices, including PCs, Macs, smartphones, and other compatible devices.

» **Read user feedback.** Beyond professional reviews, user forums can provide practical insights, such as the frequency of updates or how the software handles false positives.

PUTTING IT ALL TOGETHER

Choosing AV software is a balancing act between price, features, performance, and ease of use. By matching the tool's strengths to your unique situation, you'll find a solution that feels like a trusted ally rather than an annoying overseer.

Stay Up to Date

Updating antivirus software is crucial because it ensures protection against the newest viruses, malware, and other cyber threats, which continuously evolve. Updates include the latest virus definitions and security patches designed specifically to detect and neutralize emerging threats.

Failing to update your antivirus regularly increases vulnerability, leaving your system open to infections, ransomware, data breaches, and identity theft. Essentially, outdated antivirus software is like locking your front door but leaving your windows wide open: ineffective and risky.

WHY THIS MATTERS

» **Evolving threats.** Cybercriminals continually release new malware strains, and an outdated antivirus may not recognize these emerging threats.

» **Zero-day attacks.** As soon as security researchers discover a new exploit, vendors typically rush out updates. Missing those updates leaves you exposed.

» **Automated convenience.** Many anti-virus tools update in the background. Skipping or disabling auto-updates undermines all the good intentions of having an anti-virus solution in the first place.

KEY CONSIDERATIONS

» **Frequency.** Some AV software updates definitions daily, while others push updates every few hours.

» **Connection type.** If you're on a limited data plan, consider scheduling updates for when you're connected to Wi-Fi.

» **System reboots.** Occasional major updates might require a restart. Adjust the execution of these if you're working on time-sensitive tasks.

IMMEDIATE ACTIONS

» **Enable auto-update.** Most tools have this setting. Make sure it's on so you never miss critical patches.

» **Manual checks.** Periodically open your AV dashboard and run a manual "check for updates."

» **Stay informed.** Monitor vendor advisories or tech news for major vulnerability announcements. You might learn about updates that patch those vulnerabilities.

PUTTING IT ALL TOGETHER

Regular updates are the heartbeat of your AV/anti-malware tool. Think of them like daily vitamins; skipping them might not hurt you immediately, but it leaves you more susceptible to infections over time.

System Scans

Running regular antivirus scans identifies and removes harmful software, preventing malware from hiding unnoticed on your system. Skipping scans increases the risk of malware infections, potentially leading to stolen personal data, identity theft, system damage, or loss of essential files.

WHY THIS MATTERS

> » **Early detection.** Scans can identify suspicious files that may be hidden in downloads or system folders.
> » **Compliance.** Some workplaces require scheduled scans for regulatory purposes, especially when handling sensitive data.
> » **Peace of mind.** An occasional full scan can provide reassurance that your system remains clean and secure.

KEY CONSIDERATIONS

> » **Scan types.** Most AV solutions offer quick scans (fast but less thorough) and complete scans (slower but more comprehensive). Some also offer specialized scans for rootkits or boot sector malware.
> » **Performance impact.** Full scans can hog resources. Schedule them during off-hours or when you're less likely to need peak system performance.
> » **Frequency.** A quick scan every day or two plus a full scan weekly or monthly is a common approach.

IMMEDIATE ACTIONS

> » **Schedule scans.** Utilize your AV's built-in scheduler to run scans automatically at convenient times, such as overnight.
> » **Review scan results.** Don't just ignore alerts. Quarantines or flags should be investigated.
> » **Customize scan areas.** If you have massive external drives, consider scanning them separately to save time.

PUTTING IT ALL TOGETHER

Your software "does its rounds " by regularly scanning and checking every nook and cranny for signs of trouble. Combined with real-time monitoring, scheduled scans form a dual-layer approach, catching threats as they arrive and rooting out anything that has snuck in earlier.

Dealing with Quarantined Items

Properly handling quarantined items — by reviewing, deleting, or restoring them — is crucial because it permanently removes threats or recovers legitimate files mistakenly flagged as harmful. Ignoring quarantined items leaves your system vulnerable, potentially allowing malware to reactivate and cause further damage or data loss.

WHY THIS MATTERS

» **Preventative isolation.** Quarantine is a safe zone where suspicious files can't harm your system, like putting something in digital jail.

» **False positives.** Not every quarantined file is truly harmful; occasionally, legitimate software or files get flagged.

» **User decision.** Antivirus tools often wait for your verdict to delete, restore, or leave quarantined.

KEY CONSIDERATIONS

» **Why it's flagged.** Your AV software usually provides a reason, such as a malicious signature, suspicious behavior, or heuristic detection.

» **Risks of restoration.** Restoring a truly infected file exposes your system again. Proceed with caution.

» **Multiple opinions.** When in doubt, scanning the quarantined file with a second AV tool or an online scanner can confirm if it's genuinely malicious. In cases like this, the internet is your friend. Look up the file names or what your anti-virus program has labeled the file as and conduct an online search. In most cases, you will find thousands of other users who have experienced this detection, which can help guide your decision-making process.

IMMEDIATE ACTIONS

» **Review quarantine regularly.** Don't let items linger indefinitely. Decide whether to remove them permanently or restore them if you're confident they're safe.

» **Check file origin.** If you don't recognize the file or its source, lean toward deletion.

» **Backup before restoring.** If you must restore, consider creating a system restore point or full backup first.

PUTTING IT ALL TOGETHER

Quarantine provides a buffer between your system and potential threats. By carefully investigating flagged items, you can prevent unnecessary file deletions and reinfections of your machine.

Safe or Not?

Determining whether quarantined items are safe or dangerous is crucial because it ensures that threats are removed without mistakenly deleting legitimate files. Mishandling quarantined items can permanently delete essential data or restore malicious software, leading to system compromise and data loss.

WHY THIS MATTERS

» **Accidental deletion.** You don't want to accidentally remove a critical file due to a false positive.

» **Confirmed threats.** Ignoring a quarantine prompt could let a real threat slip through.

» **User responsibility.** AV tools aren't right 100% of the time, and they tend to err on the side of caution. What that means for you is that they will quarantine an item they're unsure of rather than assuming it is safe and letting it pass. A little detective work on your part is essential.

KEY CONSIDERATIONS

» **Signature vs. heuristic.** Some alerts originate from known malware signatures, while others are based on educated guesses about file behavior. Heuristic detections are more prone to false positives.

» **Online scanners.** Services like VirusTotal allow you to upload a suspicious file and compare it against multiple antivirus engines.

» **File context.** Did you download this file from a sketchy website? That's a red flag. Is it from a trusted source, like a reputable vendor? That's possibly a false positive.

IMMEDIATE ACTIONS

» **Scan with another tool.** A second opinion from a different AV engine can provide clarity.

» **Search for the file name.** Input the URL into a search engine such as Google. Sometimes, community or vendor forums will indicate, "Yes, xxxxx.dll is commonly flagged but safe."

» **Consult the source.** If it's from a known software vendor, check their website or support forums to see if the file is recognized as safe.

PUTTING IT ALL TOGETHER

Determining whether a quarantined item is safe requires a blend of investigative work and common sense. If in doubt, consult reliable tech communities and run multiple scans. Taking a few extra minutes to confirm can save you from losing important data or exposing your system to a malicious file.

Antivirus and anti-malware tools are essential guards for your digital front door. But the key to adequate protection goes beyond merely installing the software. You must:

» **Pick the right fit.** Evaluate the pros and cons of free versus paid solutions and identify the features that are most crucial for your specific situation.

» **Keep it current.** Enable auto-updates and schedule regular scans.

» **Use common sense.** If your software flags something, investigate; don't blindly delete or restore.

Remember, no AV suite offers bulletproof protection. Always pair these tools with solid security habits, such as safe browsing and regularly updating software. By treating your AV software as an active partner in your cybersecurity efforts, you'll stay ahead of threats and prevent your system from becoming a haven for malware. The next time you see a pop-up about a scheduled scan or an available update, don't click "Ignore." That small, simple action is crucial in your ongoing quest for layered security *(I can't let it go)*. ⚌

CHAPTER SEVEN

Phishing and Social Engineering

Scammers use fake emails, texts, and phone calls to trick people into disclosing sensitive information or clicking on malicious links. It may feel like the number of scams out there doubles every day. While this may be an exaggeration, the threats are evolving, getting more sophisticated, and harder to detect. To protect yourself, it is essential to learn how to identify these scams and threats.

Unlike traditional hacking, which relies on exploiting technical vulnerabilities, phishing and social engineering focus on manipulating individuals into providing their personal information or clicking on malicious links. These attacks can take many forms, including fraudulent emails, fake websites, and even phone calls that appear legitimate.

This chapter explains phishing and social engineering, how to identify them, and provides best practices for avoiding falling victim to these scams.

Why Phishing and Social Engineering Matter

Phishing attacks can seem innocuous at first glance, but they can lead to devastating consequences, including identity theft, financial loss, and the compromise of sensitive accounts. Cybercriminals are aware that humans are often the weakest link in cybersecurity. By exploiting human psychology, such as creating a sense of urgency or impersonating a trusted institution, they can trick people into revealing confidential information or clicking on malicious links. Recognizing and preventing these attacks is crucial for maintaining your digital security.

Check Sender Details Carefully

One of the easiest ways to identify a phishing attempt is by carefully examining the sender's email address or phone number. Cybercriminals often disguise their identity to look like a trusted company or individual, but there are usually subtle signs that give them away.

WHY THIS MATTERS

Phishing emails and texts often come from addresses that appear legitimate but with minor deviations. If something seems wrong, it's worth taking a closer look before interacting with the message. Trust your gut. If it seems suspicious, there is a good chance it's a scam.

STEP-BY-STEP GUIDE

» **Look for typos or strange characters.** Phishing emails often contain misspelled words or awkward phrasing, which serves as a red flag. Check for extra characters or a strange domain name that doesn't match the official one.

» **Examine the email address.** Even if the email is from a trusted source, check the full email address by hovering (NOT CLICKING) your mouse pointer over the sender's name. Sometimes, attackers impersonate legitimate email addresses by slightly altering the name or domain.

» **Verify with the source.** If you receive an unexpected message from someone who appears to be a trusted sender, verify the information directly with them using contact details from their official website rather than relying on the email or message itself.

BEST PRACTICES

» **Don't rely on the "From" name alone.** Always verify the sender's address, especially if the message is urgent or involves sensitive information.

» **Use caution.** Reviewing the sender's details carefully can prevent you from falling for a phishing scam. As you can see in the example below, none of the links in the email are associated with Amazon. You can check where the links will take you by hovering (NOT CLICKING) your mouse pointer over the links.

Never Click on Suspicious Links

Phishing emails and texts often include links that direct you to fake websites designed to steal your personal information. One of the easiest ways to protect yourself is to avoid clicking on suspicious links, even if they seem to come from a trusted source. By taking a few extra seconds to hover over and examine a link, you can avoid many phishing attempts that rely on misleading or malicious URLs.

WHY THIS MATTERS

Cybercriminals use disguised links to lead you to websites that look identical to the real ones, tricking you into entering login credentials or sensitive data. By checking the link before clicking, you can often avoid these traps.

STEP-BY-STEP GUIDE

» **Hover over the link.** Hover your mouse pointer over any link in an email or text message without clicking it. A small preview of the URL will usually appear in the bottom-left corner of your screen. If the URL doesn't match the official website of the organization the message claims to be from, don't click on it.

» **Manually type the URL.** If the link seems suspicious, avoid clicking it altogether. Instead, type the known website address directly into your browser to visit the site and verify any information yourself. Better yet, add the websites you use frequently to your browser's "Favorites" list so they are easily accessible.

» **Check for HTTPS.** When visiting a website, ensure the URL starts with https:// (not just http://), indicating that the website is encrypted and more secure.

BEST PRACTICES

» **Use a link scanner.** Some security software includes link scanners that verify the safety of a link before you click on it. Some websites allow you to scan a link to check if it is malicious. Try scam-detector.com or urlscan.io, for example.

» **Be suspicious of shortened URLs.** Services like bit.ly or goo.gl can disguise a malicious link. Only click on shortened URLs from trusted sources. You can expand shortened URLs using a website like expandurl.net before clicking on them.

Verify Requests for Sensitive Information

Phishing attempts often involve fake requests for sensitive information, such as login credentials, credit card details, or Social Security numbers. Reputable organizations will never request this information through insecure methods, such as email or text messages.

WHY THIS MATTERS

Scammers often pose as banks, government agencies, or popular online services to trick you into revealing personal information. By being skeptical of such requests, you can prevent sensitive data from falling into the wrong hands.

STEP-BY-STEP GUIDE

» **Look for red flags.** If an email or text message asks for sensitive information — passwords, Social Security numbers, credit card details — this is a major red flag. Most legitimate institutions will never request this information over unsecured channels, such as email or text.

» **Contact the organization directly.** If you receive a suspicious request, call the organization directly using a known phone number, found on their official website or account statements, to verify the legitimacy of the request.

» **Avoid clicking links in the request.** Instead of clicking any links in the request, go directly to the organization's website to check for security alerts or follow their official process for handling such requests.

BEST PRACTICES

» **Don't share sensitive information by email.** Even if the email looks legitimate, avoid sharing your private data via email unless you are 100 percent sure of the sender's identity.

» **Verify first.** By verifying requests for sensitive information directly with the organization, you can quickly determine whether the request is genuine or a phishing attempt.

Watch for Urgency Tactics

Phishing scammers often use urgency to trick you into taking immediate action without careful consideration. These messages may claim that your account has been compromised, your payment is overdue, or that there is a limited-time offer.

WHY THIS MATTERS

Scammers aim to create a sense of panic or urgency, which can cloud your judgment and increase the likelihood of you acting impulsively. Taking a deep breath and verifying the request through official channels can help you avoid falling into this trap.

STEP-BY-STEP GUIDE

» **Pause and evaluate.** If an email or text message claims urgent action is required, pause before acting. Take a moment to analyze the situation. Does the request seem reasonable, or is it asking for something out of the ordinary?

» **Check official communication channels.** If the message claims your account is compromised, log in to it through the official website or app rather than clicking any links in the message itself.

» **Verify the request.** Contact the organization directly using a known phone number to verify if there's truly a security issue or if it's a scam designed to create panic.

BEST PRACTICES

» **Slow down.** Scammers often try to get you to act quickly, but taking a moment to think through the situation can prevent hasty mistakes.

» **Always verify.** If you feel rushed to act, it's likely a scam. Slow down and verify the request before responding.

Phishing, Vishing, Quishing, and More

Email, text, and phone calls- modern communication makes life easier, but it also opens the door for scammers. Phishing, in all its variants, relies on social engineering to trick you into sharing personal information, clicking malicious links, or handing over money. From sophisticated corporate "whaling" schemes to quick text-message "smishing," these attacks constantly evolve. The best defense? Knowing what they look like, learning how to distinguish fake from real, and reacting swiftly when you spot a threat.

I'll dissect the most common types of phishing-style scams, offer tips to help you identify them, and guide you on the steps to take if you encounter or fall victim to such tactics.

Why Phishing Is So Common

Unlike high-tech "hacking," phishing exploits the human element: fear, urgency, and curiosity. A well-crafted email or text can bypass your defenses if you're in a hurry or your guard is down. Criminals cast wide nets (phishing) or meticulously craft personalized lures (spear phishing) to exploit weaknesses in daily routines. Recognizing their ploys is far more cost-effective than dealing with the aftermath of a compromised account or stolen identity. *However, the short answer is that phishing emails work.*

Types of Scams

Phishing scams come in multiple flavors, each targeting different channels or using varied levels of personalization. Below, I'll outline each type by name and context.

PHISHING

» **Definition.** Generic mass emails or messages that claim to be from reputable companies, such as banks or payment services.

» **Goal.** Steal login credentials, credit card numbers, or personal info.

» **Where you'll see it.** Your email inbox, sometimes social media messages, or direct messages on websites.

» **Example.** An email claiming "Your account is locked; click here to verify your password now."

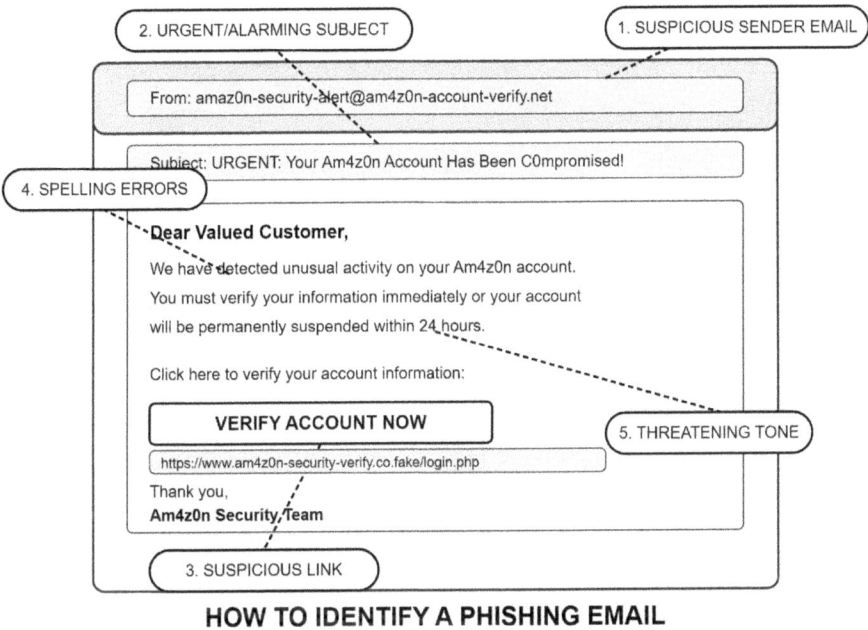

2. URGENT/ALARMING SUBJECT

1. SUSPICIOUS SENDER EMAIL

From: amaz0n-security-alert@am4z0n-account-verify.net

Subject: URGENT: Your Am4z0n Account Has Been C0mpromised!

4. SPELLING ERRORS

Dear Valued Customer,

We have detected unusual activity on your Am4z0n account.
You must verify your information immediately or your account
will be permanently suspended within 24 hours.

Click here to verify your account information:

VERIFY ACCOUNT NOW

5. THREATENING TONE

https://www.am4z0n-security-verify.co.fake/login.php

Thank you,
Am4z0n Security Team

3. SUSPICIOUS LINK

HOW TO IDENTIFY A PHISHING EMAIL

SPEAR PHISHING

- » **Definition.** A more targeted form of phishing aimed at a specific individual or small group.
- » **Goal.** Often to breach a company or high-value personal account by referencing known details about the target.
- » **Where you'll see it.** Email or direct message referencing personal information. An example might be something like this: "I saw you at the XYZ conference, can we discuss a quick opportunity?"
- » **Example.** A scammer references your boss's name and department to request a wire transfer or access to sensitive files.

VISHING

- » **Definition.** Phone-based phishing, also known as voice phishing. The attacker's call, pretending to be tech support, a bank, or even government agencies.
- » **Goal.** Coax you into giving card numbers, one-time passcodes, or remote access to your computer.

» **Where you'll see it.** Inbound phone calls, often with caller ID spoofing.

» **Example.** A call claiming, "This is your bank's fraud department. Please verify your account info immediately."

QUISHING

» **Definition.** A relatively new term, phishing via QR codes. Attackers embed malicious links in scannable QR codes.

» **Goal.** Lure you into scanning a code that leads to a counterfeit site or triggers malware.

» **Where you'll see it.** Printed flyers, emails with attached images, or "convenient" QR codes at public places.

» **Example.** A flyer promising a gift card if you "Scan Here," but the code directs you to a fake login or malicious webpage.

MISHING

» **Definition.** Phishing via SMS or text messaging, also known as mobile phishing.

» **Goal.** Trick you into clicking malicious links, sharing personal data, or calling a scam number.

» **Where you'll see it.** Text messages on your phone, often from spoofed numbers that claim to be from your bank or shipping service.

» **Example.** "Your package is delayed. Update your payment info here: [suspicious link]."

WHALING

» **Definition.** A high-stakes version of spear phishing that targets key individuals, such as executives, high-net-worth individuals, or those in critical roles within organizations.

» **Goal.** Steal large sums or sensitive data by impersonating an executive or partner.

» **Where you'll see it.** Carefully crafted emails or calls referencing company details.

» **Example.** The CEO instructs the CFO to transfer $500,000 to a new vendor account.

Real vs. Fake

Identifying a scam involves verifying details. Attackers often employ psychological tactics, such as creating a sense of urgency, instilling fear, and generating excitement, to distract you.

KEY INDICATORS

» **Generic greeting.** Phrases like "Dear Customer" instead of your actual name, especially in contexts that should be personal, for example, your bank or employer.

» **Strange email or URL.** The sender's address does not match the official domain. Hover over links to see mismatched or garbled URLs.

» **Poor grammar or spelling.** Although not always present, many phishing messages contain grammatical errors or spelling mistakes.

» **Unexpected attachment.** Be suspicious if you aren't expecting a file or invoice.

» **Urgent or threatening tone.** "Act now or your account will be suspended!"

» **Requests for confidential data.** Legitimate organizations rarely (if ever) ask for passwords, entire Social Security numbers, or verification codes via email, text, or phone.

PRO TIPS

» **Cross-check.** If in doubt, open a new browser window and manually type the official site or call the official phone line.

» **Verify through another channel.** If your CEO is emailing from an unusual address requesting a wire, contact them by phone or use an internal messaging app to confirm.

» **Search for the content.** Sometimes, an online search using phrases from suspicious emails reveals known scams.

What to Do With Threats

Immediate reaction can stop scammers in their tracks. Here's how:

» Don't click or respond.

» Delete or mark the message as spam if it's fake.

» Don't click links, open attachments, or call the phone numbers in suspicious messages.

» Notify the legitimate organization.

» Forward the suspicious email to your bank or company's official "abuse" or "security" mailbox if relevant.

» For significant email providers, you can report phishing through their built-in spam and phishing reporting tools.

» Change potentially exposed passwords.

» If you suspect you have accidentally clicked or provided incorrect information, update the credentials at risk.

» Enable two-factor authentication (2FA/MFA) on your account if you haven't already.

» Run a security scan.

» If you opened an attachment, run a full anti-malware scan.

» Keep your operating system and security software up to date.

» Alert family members or coworkers; phishing often spreads within networks if one user is fooled.

» Educate others and encourage them to verify suspicious communications.

Spam Calls and Texts

Our smartphones were once a convenient way to stay connected with family and friends. Today, they ring or buzz constantly with unwelcome spam calls and texts, often trying to trick or scare us into sharing information or money. From automated voice messages pretending to be legal authorities to AI-generated calls that sound eerily human, scammers have stepped up their game. How do you distinguish the real from the fake? I'll break down the significant types of phone and text scams, show you how caller ID can be weaponized against you, and offer practical tips to verify the authenticity of suspicious calls or messages.

Why Spam Calls and Texts Are on the Rise

With the cost of robocalling at an all-time low and technology allowing for the easy spoofing of phone numbers, scammers can cast a vast net. Voice-over-IP (VoIP) solutions further lower the barrier, letting fraudsters operate from anywhere in the world. Meanwhile, the rapid evolution of AI-driven voice tools means scamming can seem more legitimate than ever. Awareness is key; knowing the common scams and your defensive tactics will help you avoid falling for them.

How Caller ID Can Be Your Enemy

Caller ID was once our ally, letting us screen unknown or suspicious numbers. However, modern criminals manipulate this trust by spoofing.

CALL SPOOFING

- » **Definition.** Altering caller ID information so the incoming call appears to be from a legitimate source, for example, a government agency, local business, or even your own number.
- » **Tech.** Cheap software or VoIP services enable scammers to display any number they choose.
- » **Why it's effective.** Recipients see a familiar area code or recognized entity name and are more likely to pick up.

COMMON TACTICS

- » **Local spoofing.** Displaying a local area code and prefix to trick you into thinking it's a nearby call.
- » **Government or bank spoof.** Caller ID showing "IRS" or a bank's official name.
- » **Your number spoof.** In extreme cases, the call appears to come from your phone number.

AI-Generated Voice Calls

Scammers are increasingly relying on AI to generate or clone voices, sometimes using voices that are familiar to you.

» **Voice synthesis.** Tools can clone voices from small samples, replicating accents, tones, and even emotional cadences.

» **Personalization.** A call might sound like your boss or a relative asking for urgent help or money.

» **"Deepfake" calls.** Although still emerging, these can be extremely convincing, especially if you're unprepared for the possibility.

WHY THIS MATTERS

» **Emotional manipulation.** Hearing a "familiar" voice in distress lowers your guard.

» **Verification challenge.** If the call indeed sounds like someone you know, it's harder to confirm authenticity.

Fake Support Calls

Not all calls will come from a familiar voice. Some impersonate technical support or reputable businesses such as Microsoft, Apple, or Amazon in order to scam users.

» **Tech support scams.** The scammer claims your computer is infected, prompting you to install remote access software or pay for fake antivirus.

» **Bank or retailer support.** They claim there's a "suspicious charge" and request your card information or verification codes.

» **Government entity.** Posing as the IRS (tax authority) or law enforcement to threaten legal action if you don't pay immediately.

WHY THIS MATTERS

» **Urgency and fear.** "If you don't act now, your bank account will be frozen." This triggers panic responses.

» **Remote control.** In tech support scams, granting access to your device can allow scammers to install malware or steal sensitive data.

Legal Document Scam

In some spam calls or texts, you're told there's an urgent delivery of legal papers, or you're being sued or served with a warrant if you don't pay.

- » **Fake lawsuits.** "We have a civil complaint against you; call us or pay a settlement to avoid court."
- » **Fake subpoenas.** Threatening a "failure to appear" if you don't provide immediate info.
- » **Courier scams.** A text claiming a missed delivery for a "court summons" that leads to a malicious link.

WHY THIS MATTERS

- » **Fear factor.** Legal language can intimidate people into compliance.
- » **Data grab.** The scammer may prompt you to provide personal information (such as your Social Security number and birthdate) to verify your identity.

OTHER COMMON SCAMS

- » **"Grandparent" scam.** A frantic call or text claiming to be a relative in an emergency.
- » **Lottery or prize winnings.** "Congratulations, you've won a free vacation, just pay a small fee!"
- » **Debt collection.** Aggressive calls demanding immediate payment for a debt you don't recognize.

Verification Methods

These days, you may not know whether a call or message is real or not. With the surge in spoofing, AI-generated calls, and all manner of trickery, verifying authenticity can be challenging, but there are proven steps:

- » **Hang up and call back.** If it's supposedly your bank or a government agency, look up the official customer service line. Call them directly, not the number you received. Authentic organizations will never penalize you for verifying their identity.
- » **No urgent data-sharing.** Reputable entities don't demand immediate personal info or payment over the phone. If the call threatens dire consequences — such as arrest or account closure — within minutes, it's almost certainly a scam.

» **Don't trust caller ID.** Consider all caller ID info questionable. Even local or official-sounding numbers can be spoofed. Confirm whether the organization attempted to contact you through secure, official channels, such as web portal messages, official email, or letters.

» **Ask for written confirmation.** Request an emailed or mailed statement. Legitimate companies are happy to provide official documentation. Scammers usually resist giving you time to verify.

» **Cross-reference.** If a "family member" calls in distress, verify with another relative or person using known contact details.

» **Search online.** Use a search engine to look up phone numbers or suspicious phrases. Many scam phone numbers are reported online.

» **Use built-in tools and apps.** Some carriers and third-party apps (Truecaller, Nomorobo) can flag known spam numbers.

» **Update.** Keep your phone's software up to date for the latest anti-spam features.

PUTTING IT ALL TOGETHER

Phishing and social engineering attacks exploit human psychology, leveraging our natural tendencies to trust and respond quickly to perceived threats. By understanding the common tactics scammers use and adopting the best practices outlined in this chapter, you can dramatically reduce the likelihood of falling victim to these attacks. Always be cautious when sharing personal information. Scrutinize emails or texts for signs of fraud and verify any suspicious requests through official channels.

From vishing phone calls to QR code "quishing," phishing techniques thrive because they exploit the human tendency to trust. Criminals bank on you rushing to fix an alleged problem or snag a promised reward. The best countermeasure? A blend of caution and verification.

» **Learn the variants.** Phishing, spear phishing, vishing, quishing, mishing, and whaling each utilize different channels and targeting levels.

» **Spot fake vs.** real. Urgent demands, domain and sender address mismatch, grammatical errors, unusual addresses, and requests for private information are major red flags.

» **Respond appropriately.** Never click on suspicious links or share personal information. Report the attempt, change passwords, and run security checks if you suspect a breach has occurred.

Armed with knowledge, you won't just protect yourself; you'll also shield coworkers, friends, and family from falling prey to these scams. The next time you receive an "urgent request" to verify your credentials, slow down, scrutinize, and verify because, in phishing, a moment's hesitation can save you hours (or months) of recovery.

Spam calls and texts are more than just annoying; they can be downright dangerous, leading to financial loss, personal data theft, and endless stress. With caller ID spoofing and AI-generated voices, relying on superficial signs like a familiar number or voice is no longer a reliable safeguard. Instead:

» **Stay skeptical.** Don't share personal info or pay fees based on a threatening call.

» **Hang up and verify.** Check official sources, such as your bank, utility company, or government website.

» **Use tools.** Blocking, reporting, and using anti-spam apps can reduce scam attempts.

» **Spread the word.** Alert friends, family, or coworkers, especially those who may be more trusting or less tech-savvy, about emerging phone-based scams.

Combining common-sense checks with modern call-blocking tools allows you to keep these pesky, sometimes perilous, calls and texts at bay. And remember: if it feels off, it likely is. Better safe than scammed. Staying vigilant and skeptical of unsolicited requests will protect you from phishing and social engineering scams, helping to ensure that your personal and financial information remains secure. ‡

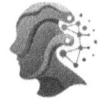

CHAPTER EIGHT

Identity Theft and Recovery

It's scary to think someone could pretend to be you, using your personal information to open lines of credit, file fraudulent tax returns, or access medical services. Identity theft can strike anyone, costing time, money, and peace of mind to fix. In this chapter, I'll break down the different types of identity theft, the signs that you might be a victim, and the best practices for preventing it. I'll also outline the step-by-step actions to take if you (or someone you know) become a target, including key contacts and an identity theft recovery checklist.

The Threat of Identity Theft

Your identity is more than just your name; it's also your Social Security number, financial accounts, medical history, and sometimes even your child's personal information. Criminals can use these details in ways that range from stealing money to committing crimes in your name. Cleaning up the mess takes significant effort and can impact your credit score, finances, and even legal standing. Understanding how identity thieves operate and taking the necessary measures when compromised helps you stay one step ahead.

What Are the Types of Identity Theft?

Identity theft isn't just about draining your bank account; different kinds of fraud target various aspects of your personal information.

FINANCIAL IDENTITY THEFT

- » **Primary goal.** Gain access to your bank accounts, credit cards, or loans.
- » **Methods.** Stolen card numbers, phishing for online banking credentials, or using your SSN to open new credit lines.

» **Impact.** Damaged credit score, unauthorized transactions, possible debt collection notices in your name.

MEDICAL IDENTITY THEFT

» **Primary goal.** Use your health insurance or medical benefits.

» **Methods.** Thieves may get your insurance policy details or health ID to obtain prescriptions, treatments, or surgeries.

» **Impact.** Inaccurate medical records (wrong blood type, allergies), denied coverage, and unexpected medical bills.

TAX IDENTITY THEFT

» **Primary goal.** File fraudulent tax returns to claim refunds.

» **Methods.** Using your SSN or ITIN before you do, changing mailing addresses or direct deposit information to intercept refund checks.

» **Impact.** Rejected tax returns, delayed legitimate refunds, IRS complications.

CRIMINAL IDENTITY THEFT

» **Primary goal.** Commit crimes under someone else's name to avoid legal consequences.

» **Methods.** A stolen driver's license or personal information is provided to law enforcement at the time of arrest.

» **Impact.** Warrants or arrests in your name, resulting in a tainted criminal record that you must clear.

CHILD IDENTITY THEFT

» **Primary goal.** Use a minor's SSN or details, which often go undetected for years.

» **Methods.** Opening lines of credit or utilities in a child's name.

» **Impact.** A tarnished credit score is often discovered only when the child applies for a student loan, job, or rental property.

Warning Signs of Identity Theft

Early detection is crucial. If you notice any of the following, investigate immediately:

» **Unexplained charges.** Bills or transactions you don't recognize on bank or credit card statements.

» **Missing mail.** Not receiving statements or tax documents could indicate that someone has changed your mailing address.

» **Credit report discrepancies.** Unknown accounts, loans, or inquiries on your credit report.

» **Collection calls.** Harassment from debt collectors for unfamiliar debts.

» **Strange medical bills.** Medical bills or insurance statements referencing treatments you never received.

» **IRS notices.** Notification that multiple tax returns were filed in your name or that IRS records show income you didn't earn.

How Can I Best Protect Myself and Others?

While identity theft can happen to anyone, several proactive steps can drastically reduce your risk.

USE ALERTING ON ACCOUNTS

» **Bank and credit card alerts.** Most institutions allow you to enable text or email notifications for transactions exceeding a certain amount or those initiated overseas.

» **Credit monitoring.** Services like Experian, TransUnion, or Equifax often have paid or free plans that alert you to changes in credit files.

» **Account statements.** Regularly review statements for suspicious charges.

PHYSICAL SECURITY

» **Sensitive documents.** Shred any documents containing personal information, such as bank statements, tax forms, and pre-approved credit offers, before discarding them.

» **Mail handling.** Consider using a locked mailbox or a P.O. box, especially if sensitive mail, such as new debit cards, has gone missing.

ONLINE HYGIENE

» **Strong, unique passwords.** Use a password manager and don't reuse passwords across accounts.

» **2FA/MFA.** Wherever possible, enable two-factor authentication (2FA) or multi-factor authentication (MFA), such as one-time codes or authenticator apps.

» **Anti-phishing practices.** Don't click on email links or share personal data with unsolicited callers or through text messages.

EDUCATE FAMILY AND CHILDREN

» **Kids.** Teach them not to share personal details, such as their birthdate or address, on social media.

» **Seniors and parents.** Help them identify scams, suspicious phone calls, or email requests for personal information.

Financial/Identity Recovery Steps

Despite precautions, identity theft can still occur. When it does, swift action is essential.

STEP-BY-STEP ACTIONS IF YOU BECOME A VICTIM

» **Contact your financial institutions.** Report stolen cards or suspicious transactions immediately. Freeze or close compromised accounts and request new account numbers.

» **Place a fraud alert or freeze on your credit files.** Notify the three major credit bureaus: Equifax, Experian, and TransUnion. A fraud alert is free and ensures that creditors must verify your identity before opening new accounts. This does not prevent creditors from accessing your credit report. A credit freeze locks your credit report, preventing new lenders from accessing it unless you lift the freeze.

» **File an identity theft report.** Visit IdentityTheft.gov (USA) or the equivalent site in your country. They'll guide you through the recovery process. Consider filing a police report, especially for significant theft or if you suspect local involvement.

» **Review all accounts and reports.** Check your credit reports for unfamiliar accounts. Audit bank statements, insurance, and utility bills for anomalies.

» **Document everything.** Keep records of phone calls, reference numbers, and the names of individuals you spoke to at financial institutions or government agencies. Write down the dates and times of all communication.

» **Change passwords and PINs.** Update logins for email accounts, banks, investment accounts, and other systems and platforms. Use fresh, unique credentials and enable multi-factor authentication.

IMPORTANT CONTACTS

» Credit bureaus (USA)
» Equifax: 800-525-6285
» Experian: 888-397-3742
» TransUnion: 800-680-7289
» Federal Trade Commission (FTC): IdentityTheft.gov
» IRS (For Tax ID Theft): 800-908-4490
» Local Law Enforcement: File a police report if necessary.

IDENTITY THEFT RECOVERY CHECKLIST

» Alert your bank/creditors ASAP.
» Set fraud alerts/credit freeze.
» File an official identity theft report.
» Check all statements and reports.
» Change all passwords/PINs.
» Stay vigilant. Monitor credit for the next 6 to 12 months and watch for further suspicious activity.

PUTTING IT ALL TOGETHER

Identity theft can be financial, medical, tax-related, or criminal, or it can even target your children's untouched credit records. Being proactive, via account alerts, shredding documents, and practicing solid online security, significantly reduces your exposure. Yet, even the most vigilant individuals can become victims. Recognizing warning signs early, such as unusual transactions or missing mail, allows you to take swift action.

If you do fall prey, document everything as you progress through the recovery process. This includes contacting financial institutions, placing fraud alerts, and filing official reports. Keep a systematic record of who you speak to and when. The IdentityTheft.gov website, designed for U.S. residents, offers a stream-lined approach; however, many other countries have similar resources available. Ultimately, the goal is to quickly shut down fraudulent activity, repair any damage to your credit or records, and fortify your defenses for the future. By staying informed and prepared, you can minimize both the likelihood and the impact of identity theft. ⁑

CHAPTER NINE

Encryption, Data Security, and Backups

WHEN YOU THINK ABOUT SECURITY, LOCKS and keys likely come to mind. In the digital realm, encryption serves a similar purpose, transforming your data into unreadable gibberish so that only someone with the correct key can unlock it. Whether you're storing sensitive files on your laptop, transferring data across the internet, or simply saving a password on your phone, encryption is the technical backbone that keeps prying eyes at bay.

In this chapter, I'll break down what encryption is, when and where to use it, and the tools available in popular operating systems like iOS, macOS, Android, ChromeOS, and Windows. I'll also address common pitfalls, such as losing your encryption keys, and examine whether third-party tools might offer better options than your built-in solutions.

Plain Text (Email)	Encryped Text (Unreadable)
Subject: Confidential Meeting Notes	Subject: U2FsdGVkX1+/5e8b9klfsdq72K/Sxmvlb=
From: jane.doe@example.com	From: R29vZGx1V2sgaGFja21uZyB0aGlzlQ==
To: john.smith@example.com	To: am9obi5zbWI0aEBleGFtcGxlmNvbQ==
[Body:]	[Body:]
Hi John,	aln Ks8dKf3J0f9klmlxD2q8Av9cW47SnH3Q==
Please keep this confidential.	**pKJZtlg2VwF7u5Xy8uL12R9a4Vd8sZq33r==**
Here are the key points from today's meeting:	0m8Ls/FjkpK3s8D4X9aPpQICl12s98Yn==
- Budget increased by 15%	jB2mX+LkmH8w4VdP3Lt4Xas7MpY7K==
- Hiring two new IT specialists	x71aZ7hlpW4q0PdH21O9b8U6JkK8=
- Discussed vulnerabilities in current infrastructure	6k8d2P0wlqj98Z2mN3a7XzE2d4H8==-
Regards, Jane	Qp72XxW8yU4m2P8qLs9b0K2j NI==

Securing Your Data with Encryption

Imagine writing a postcard: anyone can read the message simply by handling it. Encryption is like slipping that postcard into a tamper-proof envelope that only you and your intended recipient can open. In a world of data breaches, surveillance, and identity theft, encrypting your personal or professional information is often your most substantial layer of defense.

WHY THIS MATTERS

» **Fundamental concept.** Encryption is the process of encoding information so only authorized parties can read it.

» **Mathematical prowess.** Modern encryption employs complex algorithms that are nearly impossible to brute-force with current technology, assuming strong keys are chosen.

» **Everyday use.** Even if you're not aware of it, encryption is in action when you visit websites that start with "https://", access your bank app, or secure your Wi-Fi.

KEY CONSIDERATIONS

» **Symmetric vs.** asymmetric. Symmetric uses a single shared key for both encryption and decryption. Asymmetric public-key cryptography uses two keys: one public for encryption and one private for decryption.

» **Data in transit vs.** data at rest. Data in transit involves encrypting info while it moves across networks (for example, https:). Data at rest involves encrypting stored files, device hard drives, USB sticks, or cloud backups.

IMMEDIATE ACTIONS

» **Know your data.** Identify the information that would be most damaging if exposed, such as financial details and business documents.

» **Explore built-in tools.** Your OS likely offers encryption solutions for files or entire disks (more on this below).

» **Assess risk.** If you rely heavily on cloud storage, ensure that your data is encrypted, ideally with your keys.

When to Use Encryption

Encryption protects your data by turning it into unreadable code, so even if someone steals it, they can't use it. It's important to use encryption on your personal devices and networks to keep your emails, files, passwords, and financial information safe from hackers or snoops. It's your digital lock, and you'd be reckless not to use it.

WHY THIS MATTERS

» **Appropriate protection.** Not everything requires heavy encryption, but sensitive data, such as bank information, personal documents, and work files, does.

» **Legal and compliance.** Certain professions or regulations mandate encryption, for example, your healthcare records under HIPAA privacy laws.

» **Peace of mind.** Even at home, encryption prevents thieves from accessing data if they steal your laptop or phone.

KEY USE CASES

» **Full-disk encryption.** Protects your entire device if it's lost or stolen.

» **File and folder encryption.** Restricts access to selected documents or archives.

» **Cloud storage.** Some services encrypt data on their servers, but only if you hold the key does it ensure maximum privacy.

» **External drives and USBs.** Portable media is easily lost. Encryption keeps prying eyes away. There are USB drives that are encrypted by default, feature a PIN pad, and require a PIN for access (note, these can be expensive). Some will even destroy the data on them after a predetermined number of failed access attempts.

IMMEDIATE ACTIONS

» **Prioritize sensitive files.** Start with your financial or personal docs.

» **Consider device loss.** If you'd panic at the thought of someone rummaging through your phone or laptop, enable full-disk encryption.

» **Check work requirements.** If you handle company data, follow any guidelines for storing or transmitting.

Encryption Tools by Operating System

Use your operating system's built-in encryption tools, like BitLocker on Windows or FileVault on macOS. They're free, already integrated, and optimized for performance and compatibility. They're simple to enable and offer strong protection for most users.

Consider third-party tools when you need cross-platform support, extra features (like encrypted messaging or file sharing), or more control over encryption keys. Use them if you're protecting especially sensitive data or working in a regulated environment.

IOS (IPHONE, IPAD)

» **Hardware-based encryption.** When you set a passcode, the device automatically encrypts all data at rest using the Advanced Encryption Standard (AES). If you use a weak passcode, however, your data is at risk if someone can guess the code, regardless of the encryption.

MAC OS (APPLE)

» **FileVault.** Apple's full-disk encryption tool.
» **Caveat.** Ensure that you store the recovery key securely. If you lose it, you could be locked out forever.

ANDROID

» **Device encryption.** Modern Android phones (6.0 and above) typically enforce encryption by default. Some allow toggling it on and off.
» **Lock mechanism.** Ensure your device's lock (PIN, pattern, or password) is strong.

CHROME OS

» **Automatic encryption.** All user data is automatically encrypted on ChromeOS devices. Each user's data is isolated using separate encryption keys.

» **User action.** No manual setup is needed; secure your Google account with a strong password or passphrase.

WINDOWS

» **BitLocker.** Note options vary by Windows version.

» **Recovery key.** Store the recovery key in a secure location, either on OneDrive or in offline storage. You'll need it if you forget your password or if the drive's security chip resets.

WHY THIS MATTERS

» **Pre-installed solutions.** You may not need additional software if your operating system already provides encryption.

» **Ease of use.** Built-in tools are typically well-supported and user-friendly, just a few clicks away.

» **Automatic is good.** iOS and ChromeOS handle encryption behind the scenes, reducing user error.

IMMEDIATE ACTIONS

» **Check your device.** Verify the encryption status in your operating system settings.

» **Set a secure lock screen.** For mobile devices, encryption is only as strong as the passcode or biometric that protects it.

» **Backup recovery keys.** Whether it's FileVault on macOS or BitLocker on Windows, never skip the step of saving that recovery key.

Enabling Encryption Tools

Enabling your operating system's encryption protects your data if your device is lost or stolen; without it, anyone can access your files just by removing the hard drive or guessing your login. If you don't enable encryption, your info, passwords, and financial data are open to anyone with basic tech skills. It's like leaving your front door unlocked with a neon sign that says, "Come on in."

IOS

» Set a passcode in Settings > Face ID/Touch ID and Passcode.

» Use a secure alphanumeric code, not simply 4 digits.

MAC OS (FILEVAULT)

» Go to Apple menu > System Settings (Ventura) or System Preferences (earlier).

» Privacy and Security > FileVault > toggle On.

» Save your recovery key or store it in iCloud (if prompted).

ANDROID

» Check Settings > Security and Privacy > More security and Privacy > Encryption and Credentials (name may vary by manufacturer) > Encrypt phone.

» If it's not already on, follow prompts to enable.

» Ensure you use a complex PIN or password.

CHROME OS

» Encryption is on by default. Just secure your Google account credentials.

» For more control, manage local user profiles carefully.

WINDOWS

» Go to Settings > Privacy and Security > Windows Security > Device security > Data encryption.

» Or in Control Panel > BitLocker Drive Encryption.

» Turn on BitLocker for your main drive (and external drives, if desired).

» Store your recovery key safely.

The Risks of Using Encryption

Encryption is powerful, but it's not a magic wand. If you lose the key (password, PIN, or recovery code), your data will be permanently scrambled.

LOST ENCRYPTION KEYS

» **Cause.** You forgot to save your recovery key or password.

» **Fix.** Typically, there is none. Data is gone unless you somehow recall the key.

FORGOTTEN PASSWORDS OR PINS

> » **Cause.** Not using it frequently or changing it and forgetting the update.
> » **Fix.** Some systems allow partial recovery if you have a backup code.

LOST DATA

> » **Cause.** A hardware failure or accidental reformatting of an encrypted volume can render data irretrievable.
> » **Fix.** Regularly back up to an encrypted external drive or cloud service.

WHY THIS MATTERS

> » **Serious consequences.** Encryption is unforgiving if you lose your credentials.
> » **Plan.** Implementing good backup and key-management practices ensures you don't put yourself and your data at risk.
> » **Backup strategy.** Always keep a separate unencrypted (or differently encrypted) backup if you can't afford to lose the data.

IMMEDIATE ACTIONS

> » **Backup.** Keep both data backups and a secure record of encryption keys and passwords.
> » **Use a password manager.** Let it generate and securely store complex passcodes for encrypted drives or vaults.
> » **Store recovery keys.** Print them out or store them on a USB drive in a physical safe. You'll thank yourself later if you lock yourself out.

Explore Better Tools

Built-in tools that come with your OS often suffice for everyday encryption, but advanced or specialized needs might warrant third-party solutions.

VeraCrypt (Windows, macOS, Linux)

> » Open-source disk encryption tool, successor to TrueCrypt.
> » Features hidden volumes and strong encryption options.

Boxcryptor / Cryptomator (Cloud Encryption)

» Encrypt files before they sync to cloud services like OneDrive or Dropbox.

» Keeps your keys local, ensuring cloud providers can't see your data.

GnuPG (Command-Line)

» Asymmetric encryption for emails and files.

» Great for secure file sharing but has a steeper learning curve.

WHY THIS MATTERS

» **More flexibility.** Third-party tools can manage cross-platform encryption or specialized setups, for example, hidden partitions.

» **Open-source audits.** Some people prefer open-source solutions for transparency.

» **Support and maintenance.** OS vendors typically offer robust support for native solutions, which might be easier for novices.

IMMEDIATE ACTIONS

» **Evaluate complexity.** If OS-level tools do the job, stick with them.

» **Research.** If you need advanced features, such as multi-user secure containers or hidden volumes, consider reputable third-party apps.

» **Test first.** Try encrypting non-critical data to understand the workflow and confirm stability before trusting a new tool with mission-critical files.

Data Security and Backups

In today's digital age, data is more valuable than ever. From personal documents and photos to sensitive financial and health information, the data stored on our devices is integral to both our personal and professional lives. Securing that data is crucial, as its loss or theft can have severe consequences. Whether you're concerned about accidental deletion, device theft, or a cyberattack, it's essential to have a robust strategy in place to protect your data.

I will guide you through the most effective methods for safeguarding your data against loss, theft, or compromise, including regular backups, encryption, and managing outdated accounts.

What's at Stake with Data Security

Data is often referred to as the "new oil," and for good reason. Our lives are increasingly lived through a digital landscape, with personal information, work documents, banking credentials, and even our entire digital identity stored in the cloud or on physical devices. The value of that data makes it a target for cybercriminals, which is why protecting it is paramount.

» **Data loss.** From accidental deletion to hardware failure, data loss remains a persistent threat.

» **Data theft.** Cybercriminals continually seek new methods to access your data for fraudulent purposes, identity theft, or more severe consequences.

» **Privacy concerns.** The more data we store, the higher the risk of exposing sensitive information.

Fortunately, with the proper precautions in place, you can ensure that your data remains safe and secure. This chapter outlines the essential steps to take for backing up, encrypting, and managing your data.

Back Up Important Files Regularly

One of the simplest yet most effective ways to protect your data is by creating regular backups. Backups ensure that if your device fails, is lost, or is attacked, you won't lose all your important information.

WHY THIS MATTERS

No system is infallible, and data loss can occur for various reasons, including hardware failure, accidental deletion, or ransomware attacks. Backing up your data is your safety net, ensuring you can restore it even in the event of a catastrophe.

STEP-BY-STEP GUIDE

Choose a backup method

» **External hard drives.** Store copies of important files on external drives, such as USB drives or portable hard drives. These are useful for offline backups and can be stored securely at home or in another location.

» **Cloud services.** Utilize cloud storage providers such as Google

Drive, Dropbox, or OneDrive for seamless access to your files from any device.

Set up automatic backups

WINDOWS

- » Go to Control Panel > Backup and Restore (Windows 7).
- » Or enable File History.
- » Or copy essential files to OneDrive for automatic backup and syncing..

MAC

- » Use an external drive with Time Machine to back up data.
- » Create regular backups.
- » Whether you use external drives or cloud services, ensure that backups are created regularly. Set automatic backups whenever possible, and manually back up important files as needed.

BEST PRACTICES

- » **Follow the 3-2-1 Rule.**
  - » 3 copies of your data
  - » 2 different types of media
  - » 1 copy stored off-site
- » **Maintain three copies of your data.** This includes the original data and at least two copies.
- » **Store your data on two different types of media for redundancy.** For example, an external hard drive and a large-capacity USB flash drive.
- » **Keep at least one copy off-site.** To ensure data safety, have one backup copy stored in an off-site location, separate from your primary data and on-site backups.
- » **Test your backups.** Periodically check your backup files to make sure they are accessible and intact.

Regular backups minimize the risk of losing your important data, ensuring that even in the worst-case scenario, you can recover your files quickly.

Delete Old Accounts

We all accumulate online accounts over time, some we may no longer use or need. Leaving old accounts open can expose your data to breaches or hacking attempts, especially if you've reused passwords or neglected to update security settings. By deleting old or unused accounts, you reduce the number of digital entry points hackers could exploit.

WHY THIS MATTERS

Every account you create is an additional place where your data is stored, and inactive accounts are especially vulnerable. Deleting accounts you no longer use reduces the number of places where your data is exposed to potential risks.

STEP-BY-STEP GUIDE

Review your online accounts

- » Make a list of all online services you use, including social media, shopping, banking, and subscription-based services.
- » Periodically review this list to identify accounts you no longer need or use.

Close unused accounts

- » Most services allow you to delete your account through their settings or privacy section. Check the service's help section or contact support if you can't find the option.

Delete personal information

- » If you can't delete the account, consider at least removing personal information (such as your address or payment details) to minimize exposure.

Use account deletion services

- » If you have many old accounts, consider using services such as JustDelete.me or AccountKiller, which provide direct links and information on deleting accounts from various websites.

BEST PRACTICES

- » **Review permissions.** Before deleting accounts, review the permissions granted to them and remove anything that might compromise your data security.

» **Delete for security.** For accounts that store sensitive data (such as banking or healthcare), ensure they are adequately deleted, including removing your information from the site.

PUTTING IT ALL TOGETHER

Encryption is the lock on your digital front door, ensuring that even if thieves get physical or network access, they can't read your files without the correct key. Native solutions, like FileVault, BitLocker, iOS hardware encryption, or ChromeOS default encryption, make it simpler than ever to protect devices at rest. Meanwhile, advanced or specialized encryption tools provide more granular control for those who need it.

However, with great power comes great responsibility: lose the key, lose the data. Maintaining backups, storing recovery keys securely, and utilizing robust password management are essential to prevent locking yourself out. Evaluate which method and scope of encryption best suits your personal or professional data needs. By combining strong encryption practices with prudent key management, you can make your data an impenetrable fortress.

Data security and backups are essential to maintaining the privacy and integrity of your personal information. By regularly backing up your essential files, encrypting sensitive data, and deleting old accounts, you take proactive steps to protect your digital life. These measures significantly reduce the risk of data loss and theft, ensuring that even if your devices or accounts are compromised, you have safeguards in place to recover and secure your information.

It's easy to overlook the importance of data protection until it's too late, but by following these simple steps now, you can ensure that your data remains safe for years to come. Stay vigilant about your data, back it up regularly, and maintain a secure digital footprint. ⁑

CHAPTER TEN

Application and Browser Security

Apps on your phone, tablet, and computer can be weak points if not managed correctly. The average person now has dozens, if not hundreds, of apps installed across their devices, from productivity tools and social media platforms to banking and shopping apps. While these apps offer convenience, entertainment, and functionality, they can also introduce significant security risks if not managed carefully. Some apps contain vulnerabilities that can expose your personal information, allow unauthorized access, or even infect your device with malware.

Fortunately, securing your apps doesn't have to be complicated. With a few simple steps, you can minimize the risks associated with apps and ensure your devices remain secure. This chapter will guide you through best practices for app security, including where to download apps from, how to manage app permissions, and how to evaluate free apps for potential privacy concerns. I'll cover steps you can use to prevent these risks and ensure that your apps are as secure as possible.

Challenges of App Security

Apps are integral to modern life, but they also serve as a standard entry point for malicious software (malware) and unauthorized data collection. Because apps often request or require broad access to sensitive features on your device, such as your documents, pictures, camera, microphone, contacts, and location, they can be a goldmine for hackers if not properly secured.

Here are a few reasons why app security is crucial:

> » **Permissions and data access.** Many apps request access to data or
> features you might not expect, such as your contacts, messages, or even
> location data. As an example, the following points give a breakdown of

the Privacy Policy for a very popular social media app. I have replaced the name of the app with XXX.

» **Biometric data collection.** XXX explicitly states it may collect biometric identifiers (faceprints, voiceprints). Even with consent, handing over your biometric data to a company with extensive international ties is risky.

» **Clipboard access.** The app collects information stored in your device's clipboard. This means anything copied, potentially sensitive information, could end up with XXX, intentionally or not.

» **Cross-device and off-app tracking.** XXX tracks your activity across multiple devices and other websites or apps, associating it directly with your XXX account. Your online behavior outside XXX isn't off-limits.

» **Extensive metadata and geolocation tracking.** Although precise GPS tracking is limited in newer versions, XXX still collects extensive metadata and approximate location data via IP addresses and SIM cards, revealing far more about your daily routine than you might expect.

» **International data transfer.** XXX transmits user data internationally, potentially subjecting your information to jurisdictions with weaker privacy protections or differing standards on governmental access to personal data.

Do you think an app designed for entertainment purposes needs or should have this type of extensive access to your life? Read the Privacy Policies and Terms of Service carefully and think carefully before installing an application. Regardless of how many security measures you deploy, they are meaningless if you voluntarily share all your data with an application.

» **Malware and exploits.** Even seemingly benign apps can contain malware or be vulnerable to exploits that could compromise your device.

» **Privacy concerns.** Free apps often collect data about you, which they may sell to third parties or use for targeted advertising.

Only Download Apps from Official Stores

The first step in ensuring the safety of apps on your devices is to download them from trusted sources. Official app stores, such as Google Play for Android devices, the Apple App Store for iOS, and the Microsoft Store for Windows, have security measures to prevent malicious apps from being published. Yes, sometimes bad apps slip through the cracks, but these are your safest bet overall.

WHY THIS MATTERS

Sideloading apps — downloading apps from third-party websites or unofficial sources — exposes you to significant risks, as these apps may contain malware, spyware, or other malicious code. Official app stores vet apps for security, reducing the likelihood of downloading a harmful app. These stores implement basic app vetting processes.

STEP-BY-STEP GUIDE

» **Download from official stores only.**
  » Android: Use the Google Play Store.
  » iOS: Use the Apple App Store.
  » Windows: Use the Microsoft Store.
  » Linux: If you're using Linux, you probably aren't reading this book.
» **Check app developer information.** Before downloading, ensure the app developer is reputable. Look for developer information and read reviews from other users.
» **Avoid third-party app stores.** Steer clear of app stores or websites that aren't official sources. These stores may host apps that have not been vetted for security.

BEST PRACTICES

» **Regularly review app permissions.** Even apps from trusted sources can request more permissions than necessary. Regularly review which apps have access to your sensitive data. For example, does your banking app require your location? Maybe if you're looking for an ATM, but in general, no. These types of permissions can often be turned off until needed or be set to "Only while using this app.".

» **Pay attention to app updates.** Only download updates for apps from official sources to avoid accidentally installing malicious updates.

Sticking to official app stores minimizes the chances of downloading malware or compromised apps, which is your first line of defense.

Check App Permissions

Modern apps often require access to various parts of your device, including your camera, microphone, contacts, and location. While some permissions are necessary for the app to function — like when a GPS app requires access to your location — others may not be as crucial and can compromise your privacy if left unchecked.

Managing app permissions ensures that only the apps that require access to your data or device features have permission to do so, thereby minimizing the risk of unauthorized data sharing.

WHY THIS MATTERS

Apps can request far-reaching permissions that may not be necessary for their core function. For example, a flashlight app doesn't need access to your contacts, and a game app doesn't need to track your location. Limiting unnecessary permissions reduces the potential for data misuse. There may come a time when you will need to decline installing an application because you feel it is requesting excessive access. Better safe than sorry, and chances are there is another app that does the same thing and does it more securely.

STEP-BY-STEP GUIDE

Review app permissions

» Android: Go to Settings > Apps > Special app access (note that this location may vary depending on your Android version).

» iOS: Go to Settings > Privacy and Security (note that this may vary depending on your iOS version).

Review the permissions for each app

» Disable unnecessary permissions. For apps that request permissions you don't think are necessary (such as location access for a calculator app), disable them. On Android and iOS, you can easily toggle off unnecessary permissions, such as access to the camera or microphone.

» Periodic review. Review the permissions for your apps every few months. Some apps may request new permissions over time, which you can choose to deny or enable.

BEST PRACTICES

» **Principle of least privilege.** Only allow apps to access the features they need to function.

» **Use app permission alerts.** On some phones, you can enable alerts that notify you when an app requests sensitive permission.

Be Wary of Free Apps

"If you're not paying for the product, you are the product." There are other similar sayings, but the point is that nothing is free.

Free apps may seem like a great deal, but many come with hidden costs: often, your data is the price. Many free apps collect your personal information, which is then sold to advertisers or third parties. Therefore, it is essential to assess the privacy policies and data collection practices of any app you download, whether free or paid.

WHY THIS MATTERS

While paying for an app doesn't guarantee privacy, free apps are more likely to engage in data mining and targeted advertising. This can result in the sharing, selling, or use of your data.

STEP-BY-STEP GUIDE

Read privacy policies

» **Data collection.** Before downloading or using a free app, check its privacy policy to see what data it collects and how that data is used.

» **Third-party sharing.** Pay attention to sections that mention third-party sharing and data retention policies.

» **See where your data is stored.** If it is outside the United States, it may be subject to the laws of the country in which it is located. In general, this should be a red flag.

<u>Check for in-app ads</u>

- » **Many free apps show in-app ads.** Be cautious, as these ads might collect data or redirect you to malicious websites.
- » **Read app reviews.** Check reviews and ratings in the app store. Look for feedback from other users about privacy concerns or potential security risks.

BEST PRACTICES

- » **Limit free apps.** Only download free apps from trusted developers and companies that are transparent about their data usage.
- » **Use paid alternatives.** When possible, opt for paid apps that don't rely on monetizing your data through ads or other means. Paid apps are generally less likely to engage in intrusive data collection practices.
- » **Due diligence.** Before installing software or agreeing to a Privacy Policy or Terms of Service agreement, conduct your due diligence. They can't get what you don't agree to give them when you hit "Accept."

The Fine Print, Simplified

Software companies traditionally have lengthy Privacy Policies and Terms of Service, and they know that 99 percent of people will click "Accept" and install the app. There is a quick way to read them and understand what you're agreeing to without having to read the entire document.

Open the policy you want to review and copy the entire thing to your clipboard. Now, open your favorite AI website. This could be ChatGPT, Claude, CoPilot, or another similar tool. Insert the following prompt:

"Review the following Privacy Policy (or Terms of Service) and let me know what I should be concerned about. List the top five points that should stop me from installing this app and put them in bullet points."

Paste the policy under the prompt you just typed and hit enter. It will provide a summary of the top five reasons not to install the application. You can alter the prompt as needed to achieve the desired response.

Being cautious with free apps can help you avoid privacy issues and ensure that you are not unknowingly compromising your personal information.

Browser Security

A web browser isn't just another app; it's your primary gateway to the internet. Every website you visit, every online form you fill out, and every ad you click on runs through your browser. While operating system or application security patches are vital, a neglected or outdated browser can open a direct line for attackers to steal data or compromise your system. Beyond security, privacy also plays a role. Trackers, ads, and cookies can follow your browsing habits, build detailed profiles, and reduce your sense of control over personal information.

Next, I'll explore why browser security differs from that of other software, how to keep each major browser up to date, and what you need to know about extensions and plugins. I've selected some of the most popular browsers to review: Safari, Chrome, Edge, Firefox, and DuckDuckGo. I'll also delve into privacy topics, such as tracking and cookies, what they are, how to manage them, and how each browser collects (or doesn't collect) your data.

Browser Vulnerabilities

Unlike most apps on your system, your browser constantly loads content from websites you may or may not trust. Malicious advertising, also known as "malvertising," or sophisticated phishing sites can quickly exploit any unpatched vulnerability. Even when your operating system is fully updated, a single vulnerability in your browser can allow an attacker to gain unauthorized access. Moreover, issues such as excessive tracking scripts or poorly secured cookies can gradually erode your privacy, often without you even being aware of it.

Different Aspects of Security

Browser security is different from your OS or application security and knowing the difference will help make sure you are protected.

DIFFERENCES

» OS security protects the entire system (hardware, memory, files). It's the foundation.

» Browser security (websites, scripts, plugins) protects your online activity.

» Application security protects specific software (email, MS Office) from exploitation.

SIMILARITIES

» All aim to prevent unauthorized access, data leaks, and system compromise.

» All need regular updates, strong permissions, and safe configurations.

Why address them individually? Because each has unique threats and entry points. A flaw in one layer — your browser, for example — can be used to bypass the others. Ignoring one weakens your entire security chain.

WHY THIS MATTERS

» **Ever-changing web content.** Each new site you visit may load scripts, ads, or media files, some of which could be malicious if your browser is unprotected or outdated.

» **User interaction.** Users frequently click pop-ups, grant permission requests, or install add-ons that can compromise security.

» **Mixed trust levels.** Your browser must handle both trustworthy sites (banking, email) and questionable sources — including links you click in unknown emails — which raises the stakes.

KEY CONSIDERATIONS

» **Sandboxing.** Modern browsers run site processes in isolation; however, exploits can still escape these sandboxes if the browser is outdated or misconfigured.

» **Phishing and social engineering.** Attackers often rely on tricking the user rather than relying solely on technical flaws. A "safe" browser can't protect you if you unknowingly install malicious extensions or click "Allow" everywhere.

IMMEDIATE ACTIONS

» **Stay educated.** Recognize that your browser is a prime target.

» **Reduce risky behavior.** Be cautious of suspicious pop-ups, especially those that urge you to install software or sign in with personal details.

» **Harden settings.** Disable unnecessary features and plug-ins and regularly review your privacy settings.

How Do I Keep My Browser Up to Date?

Keeping your browser up to date is crucial because updates fix security flaws, improve performance, and support new web technologies. If you don't update, hackers can exploit known vulnerabilities to steal data, install malware, or hijack your browsing activity. An outdated browser is like leaving the keys in your unlocked car.

SAFARI (MAC OS)

» **System updates.** Safari updates are bundled with macOS.

» **Check system settings.** On macOS Ventura or later: Apple Menu > System Settings > General > Software Update.

» **On older macOS versions.** Apple Menu > System Preferences > Software Update.

» **Automatic updates.** Toggle on "Automatically keep my Mac up to date." Safari will update with each macOS patch.

GOOGLE CHROME

» **Auto-update by default.** Chrome usually updates itself silently in the background.

» **Manual check.** Select the three-dot menu > Help > About Google Chrome. Chrome will download updates if available.

» **Restart to apply.** You may need to relaunch Chrome for the new version to take effect.

MICROSOFT EDGE

» **Similar mechanism to Chrome.** Edge auto-updates under the hood.

» **Manual check.** Three-dot menu > Help and feedback > About Microsoft Edge. If an update is pending, install and then relaunch.

» **Windows Update.** Keeping Windows up to date also ensures that Edge is updated.

MOZILLA FIREFOX

- » **Auto-update by default.** Firefox typically updates itself in the background.
- » **Manual check.** Three-line menu > Help > About Firefox. Firefox automatically checks for updates and applies them.
- » **Restart if prompted.** A quick relaunch often finalizes the update.

DUCKDUCKGO

- » **Primarily mobile.** Update on iOS or Android via the App Store or Google Play.
- » **Automatic updates.** If your phone or tablet is set to auto-update apps, DuckDuckGo Browser should stay current.
- » **Check version.** In the App Store or Play Store, look for pending updates if you suspect your app is outdated.

WHY THIS MATTERS

- » **Frequent patches.** Each browser has a rapid release cycle to fix security bugs. Missing even one update leaves known vulnerabilities open.
- » **Auto vs.** manual. By default, most browsers now auto-update. Periodic manual checks ensure you're never left behind.

IMMEDIATE ACTIONS

- » **Verify auto-updates.** Double-check that each browser's auto-update feature is enabled and functioning correctly.
- » **Restart.** Don't ignore prompts to restart after an update; they're there for a reason.
- » **Check once a month.** Perform a quick manual version check to confirm you're on the latest build.

What About Extensions and Plugins?

Extensions, such as ad blockers or password managers, and plugins (specialized video players or PDF readers) can enhance your browsing experience or streamline tasks. However, they also introduce risks if they come from untrustworthy sources or have excessive permissions.

WHERE TO SAFELY GET BROWSERS

Avoid random websites. If an extension is only available from an obscure site, think twice. Be sure to download browsers only from official browser stores:

- » **Safari.** App Store (macOS)
- » **Chrome.** Chrome Web Store
- » **Edge.** Microsoft Edge Add-ons
- » **Firefox.** Mozilla Add-ons

ARE EXTENSIONS SAFE?

- » **Vet them.** Even legitimate stores occasionally host malicious or low-quality extensions.
- » **Check ratings and reviews.** Look for large user bases and solid feedback.
- » **Permissions.** Be cautious if a plugin requests full access to all your site data when it shouldn't need it.

Managing Permissions

- » **Safari.** Safari > Settings > Extensions
- » **Chrome.** Three-dot menu > Extensions
- » **Edge.** Three-dot menu > Extensions
- » **Firefox.** Three-line menu > Add-ons and themes > Extensions
- » **DuckDuckGo.** Primarily minimal by design, with fewer extension options. Manage any integrated features via its settings.

WHY THIS MATTERS

- » **Data access.** A malicious extension can monitor your browsing activity, inject ads, or even steal your passwords.
- » **Supply chain attacks.** Sometimes, a trustworthy extension is bought out by shady actors and updated with malicious code.
- » **Performance.** Bloated extensions can slow down page loads or cause crashes.

IMMEDIATE ACTIONS

- » **Audit add-ons.** Remove or disable anything you don't frequently use.

» **Limit permissions.** If the extension or plugin allows you to restrict site access, do so.

» **Keep updated.** Legitimate developers release updates to fix bugs and patch security holes.

Privacy, Tracking, and Cookies

Security isn't the only concern: online tracking and data collection have turned into significant privacy headaches. Many websites rely on cookies to remember your preferences or sign-in status, but they can also be used (or misused) for extensive tracking across multiple sites.

Cookies and Your Privacy

Cookies are small text files placed on your browser by websites. They store information like your login status, site preferences, or tracking IDs.

» **First-party vs.** third-party. First-party cookies originate from the site you're visiting ("remembering your shopping cart"). Third-party cookies often originate from advertisers or analytics services, enabling cross-site tracking. They are sometimes referred to as "Persistent Cookies."

» **Privacy impact.** Advertisers can piece together your browsing habits, build a profile of your interests, and serve targeted ads. Websites might share or sell that data, leading to unwanted data collection.

MANAGING COOKIES

Browser-specific settings

» **Safari.** Safari > Settings > Privacy

» **Chrome.** Settings > Privacy and security > Cookies and other site data

» **Edge.** Settings > Cookies and site permissions

» **Firefox.** Three-line menu > Settings > Privacy and Security > Cookies and Site Data

» **DuckDuckGo.** Typically blocks third-party trackers by default; adjust "Privacy" settings in the app.

<u>Block or clear</u>

» **Block.** Choose to block third-party cookies, which are the main culprits for cross-site tracking.

» **Clear.** Periodically clear cookies if you want a fresh start or suspect questionable trackers. Please note that this may require you to reauthenticate on all websites you use regularly.

<u>Site-by-site control</u>

» **Control.** Some browsers allow you to manage cookies on a per-domain basis. This is handy if you want to keep login cookies but block trackers.

Comparing Privacy Levels Across Browsers

While all major browsers support essential security features, such as sandboxing and auto-updates, their privacy philosophies differ. Here's a quick snapshot:

SAFARI (MAC OS)

» **Default tracking.** Intelligent Tracking Prevention (ITP) blocks many third-party trackers.

» **Data collection.** Apple generally collects minimal user data, focusing on on-device processing for personalization.

» **Privacy focus.** High. Apple markets privacy as a selling point. However, as with everything else, *due diligence is necessary.*

GOOGLE CHROME

» **Default tracking.** Chrome is Google's primary revenue channel for ads, so it doesn't aggressively block third-party cookies by default.

» **Data collection.** Google collects usage stats and sync data unless you opt out.

» **Privacy focus.** Medium. Chrome can be customized for better privacy, but the defaults are generally ad-friendly.

MICROSOFT EDGE

» **Default tracking.** Edge uses "Tracking prevention" levels (Basic, Balanced, Strict).

» **Data collection.** Microsoft collects usage data for product improvement, but you can opt out of some data collection.

» **Privacy focus.** Medium-High. Stricter settings can significantly reduce tracking, although Microsoft does collect telemetry.

MOZILLA FIREFOX

» **Default tracking.** Enhanced Tracking Protection blocks trackers, crypto-miners, and fingerprinting by default in "Standard" mode.

» **Data collection.** Mozilla collects minimal telemetry and is transparent; you can easily opt out.

» **Privacy focus.** High. Firefox is an open-source browser that consistently adds new anti-tracking features.

DUCKDUCKGO

» **Default tracking.** DuckDuckGo's standalone browser automatically blocks third-party trackers.

» **Data collection.** Claims zero data collection or user profiling.

» **Privacy focus.** Very High. DuckDuckGo is built around privacy by design.

PUTTING IT ALL TOGETHER

Apps are powerful tools that enhance our productivity, entertainment, and everyday lives. However, they can also pose significant risks to privacy and security if not used properly. By following these steps and downloading only from trusted sources, managing app permissions, and exercising caution with free apps, you can significantly reduce the risks associated with app usage.

Remember, the key to app security is balance: you want to enjoy the benefits of modern apps without sacrificing your privacy. By staying informed, regularly reviewing app permissions, and making informed choices about the apps you use, you can help ensure that your devices remain secure while still enjoying the convenience that apps provide.

By taking these steps now, you can safeguard your data, prevent unnecessary data collection, and minimize the risk of downloading potentially harmful apps.

With these app security practices in place, you'll be well on your way to maintaining a safe and secure digital environment.

Browser security stands at the intersection of safe software practices and privacy considerations. Even with a fully patched operating system, an outdated browser or shady extension can let malware slip in. Meanwhile, online trackers and advertising networks rely on cookies and cross-site scripts to build profiles of your behavior.

» **Stay updated.** Safari, Chrome, Edge, Firefox, and DuckDuckGo all have auto-update mechanisms. Verify they're enabled and remember to relaunch after updates.

» **Extensions.** Install only from official stores, check permissions, and remove any unused extensions.

» **Privacy and cookies.** Know how cookies work, consider blocking third-party cookies, and manage them on a site-by-site basis.

» **Choose your browser wisely.** Each browser has a distinct approach to privacy. While Chrome and Edge offer robust performance with moderate tracking controls, Safari, Firefox, and DuckDuckGo take it a step further by minimizing data collection and blocking trackers.

By combining smart security habits, such as disabling dangerous add-ons and keeping up with patches, with robust privacy settings, you can transform your browser from a potential security vulnerability into a secure, privacy-respecting gateway to the online world. ⚡

CHAPTER ELEVEN

Social Media Privacy and Safety

SOCIAL MEDIA HAS BECOME AN INTEGRAL part of daily life, whether it's catching up with friends, sharing photos, or discussing current events. But each post you share can reveal more about you than you might expect. Criminals, scammers, and even over-curious acquaintances can exploit the details you provide. Is social media safe? That depends on your habits, privacy settings, and what you share, as well as with whom.

This chapter breaks down the key risks associated with social platforms, highlights why and how to limit oversharing, and offers step-by-step guidance on managing your privacy settings across prominent social media networks. Our goal is to help you continue enjoying online communities while reducing vulnerabilities to scams, stalking, identity theft, or future regrets.

Is Social Media Safe?

Once a novelty, social media now functions as a digital public square. However, with billions of users, it also serves as a treasure trove for data miners, scammers, and potential predators. A single overshared post can reveal your location, financial situation, or personal details, all of which can be misused. Understanding these pitfalls and how to mitigate them is crucial for maintaining both digital and physical security.

WHY THIS MATTERS

» **Massive reach.** Posts can quickly travel beyond your immediate circle if your privacy settings are lax or if friends reshare your content.

» **Platform changes.** Networks like Facebook, Instagram, Twitter (X), and TikTok routinely update their features, sometimes altering default privacy settings.

» **Human factor.** Even strong platform policies can't protect you if you knowingly (or unknowingly) post sensitive info or accept friend requests from strangers.

KEY POINTS

» **Default visibility.** Many platforms encourage open sharing to build engagement, so you must manually adjust privacy settings.

» **Unpredictable enforcement.** While harassment or scam posts are often reported, moderation can be inconsistent.

» **Archiving and screenshots.** Once content is online, it can be archived or copied even if you later delete it.

IMMEDIATE ACTIONS

» **Review your profile as a stranger.** Many platforms allow a "View as Public" option.

» **Customize privacy.** Switch from "Public" to "Friends Only" or "Connections Only."

» **Think before you click.** Apply skepticism before donating to suspicious fundraisers or opening unknown links.

What Are the Risks?

Using social media can be fun. It can also expose you to privacy invasion, identity theft, scams, cyberbullying, and misinformation. Sharing too much personal info can make you a target for hackers or stalkers. What you post can also hurt your reputation or job prospects.

SCAMS (PHONY FUNDRAISING, DISASTER RELIEF EXPLOITS)

» **Fake charities.** These exploit emotional triggers after natural disasters or crises.

» **Phishing tactics.** Urgent requests that push you to enter personal info or send money quickly.

STALKERS (PHYSICAL SECURITY)

» **Location data.** Check-ins or geotags can reveal where you live, work, or spend free time.

» **Personal threat.** A persistent stalker may track you in real life if you inadvertently share enough details.

THIEVES (BURGLARS)

» **Vacation posts.** Publicly announcing you're away can be a burglar's dream invitation.

» **Property posts.** Posting photos of expensive gadgets or cars can draw the attention of thieves.

CYBERCRIMINALS (ACCOUNT/FINANCIAL THEFT)

» **Account takeovers.** Weak passwords or easily guessed security questions, such as your pet's name or high school's mascot, can lead to hijacked profiles.

» **Fraudulent posts.** Hackers posing as you to trick friends into sending money or personal information.

FAKE JOB POSTINGS

» **Data harvesting.** Collecting resumes for identity theft.

» **Advance fee scams.** "Pay for your training materials" hoaxes or "check processing" cons.

OTHERS (CATFISHING, EXTORTION)

» **Catfishing.** Someone impersonating an entirely different person to form relationships for emotional or financial gain.

» **Extortion.** Threatening to leak private images or messages unless you pay or comply with demands.

IMMEDIATE ACTIONS

» **Verify causes and people.** Double-check authenticity before donating or responding to emotional pleas.

» **Limit personal info.** Keep location tags, birthdates, and personal identifiers private.

» **Stay alert.** If a friend posts or messages something that seems "off," it may be from a hacked account. Call them to confirm if the post is legitimate first.

Limit What You Share

If you wouldn't share something in a room full of strangers, don't share it on the internet. This principle is your first line of defense against oversharing. Here are some examples:

» **Vacations.** Post about the trip once you're home, rather than announcing you're away in real time.

» **Addresses and license plates.** Crop or blur these in photos.

» **Kids and pets' names.** Common security question answers; avoid revealing them publicly.

» **Birthdays and anniversaries.** Identity thieves can use these special days to guess passwords or answer "secret questions."

WHY THIS MATTERS

» **Permanent records.** Data can remain in hidden archives or be reposted by others.

» **Accumulated clues.** Even if each post seems harmless, together, they can form a detailed profile of your life.

IMMEDIATE ACTIONS

» **Check for background details.** Ensure that your home address, financial documents, or sensitive items aren't visible.

» **Delay posting.** If location-based content is essential, consider time-shifting to after you've left.

» **Set boundaries with friends and family.** Politely request that they avoid tagging your location or posting your updates without your permission. Alternatively, you can prevent yourself from being tagged in other people's posts by adjusting your settings.

Limit Whom You Share With

Even if you share something personal, you can restrict its audience to specific groups or lists of trusted contacts.

TIPS

» **Private vs. public.** On most platforms, your default sharing option can be changed to "Private," "Friends Only," or "Connections Only."

» **Friend lists and groups.** Facebook and other sites let you create custom lists.

» **Follow and connection approvals.** Keeping your account private — which is an option on some apps such as Instagram and TikTok — or selectively accepting friend requests helps limit unknown onlookers.

WHY THIS MATTERS

» **Mitigates oversharing.** Even personal posts might not be considered too sensitive if only your inner circle sees them.

» **Protects workplace and personal boundaries.** Keep professional and personal contacts in separate categories, if possible.

» **Reduces "leaking" information.** While not foolproof, fewer eyes on your posts mean fewer chances for them to be reshared.

IMMEDIATE ACTIONS

» **Trim your network.** Remove connections you don't know or trust. Do you really have 1,000 "friends?"

» **Experiment with lists.** Utilize friend lists to target or exclude specific individuals for particular posts.

» **Review privacy options.** Some platforms allow you to fine-tune post visibility; learn how to do it effectively.

The Internet Is Forever

Even if you delete a post, screenshots or cached versions may still be available. Don't share something today that you'll regret tomorrow or forever.

WHY THIS MATTERS

- » **Future employment.** Potential employers often research social media profiles. Controversial or offensive content can haunt you for a lifetime.
- » **Legal ramifications.** Posting private or privileged information publicly can have serious consequences.
- » **Personal reputational damage.** Emotional rants or personal confessions might resurface at an awkward time.

IMMEDIATE ACTIONS

- » **Think twice.** Pause before posting in anger or haste.
- » **Use private channels.** Vent in a private message or keep a personal journal instead of blasting public updates.
- » **Regular cleanup.** Periodically review old posts for outdated or risky material. Platforms like Facebook offer "Limit Past Posts" or archiving features.

Managing Your Privacy Settings on Major Social Media Platforms

Beyond general advice, each social platform has its privacy controls. Please note that platforms frequently update their menus, so some of these items may have moved or been renamed by the time you read this. These items are listed here for general reference. Here are some highlights to help you safeguard your personal information:

FACEBOOK

Privacy checkup

- » Go to Settings and Privacy > Privacy Checkup for a guided review of who can see your posts, personal info, and apps.

Adjust post visibility

- » Under Settings > Privacy Center > Common privacy settings > Manage audience settings > Posts and stories, set your default post visibility to Friends, Friends Except… or Only Me.

Limit old posts

» A one-click option lets you retroactively change all public or "friends of friends" posts to "Friends Only."

Profile info

» Decide whether to hide or limit details, such as birthday, relationship status, and contact information, to "Only Me" or "Friends."

Timeline and tagging

» Enable "Review tags people add to your posts before the tags appear on Facebook?" to approve or reject them before they appear on your timeline.

INSTAGRAM

Switch to a private account

» Profile > More (three-line menu) > Settings > Account Privacy > toggle Private Account.

Close friends

» Create a "Close Friends" list for sharing more personal Stories with a select group.

Manage tags

» More > Tags and mentions > Who can tag you. Control who can tag you in posts.

» Use the Tag Controls to approve tags before they go live.

Location tags

» Turn off location tagging or only use it selectively.

LINKEDIN

Who can see your profile and network

» Profile > Settings and Privacy > Visibility. Adjust what your connections can see.

» Public Profile

» Profile > Settings and Privacy > Visibility. You can decide how much (if any) of your profile is visible to the public search results.

Connections vs. followers

» Set whether people must request to connect or if they can follow you.

Data sharing with third parties

» Profile > Settings and Privacy > Advertising data. LinkedIn can share your data with third parties for ads; turn this off under Advertising data.

TWITTER (X)

Protect your tweets

» More > Settings > Privacy and Safety > Audience, media, and tagging > Enable "Protect your posts" so only people who follow you can see them.

Location info

» More > Settings > Privacy and Safety > Data sharing and personalization > Location information.

» More > Settings > Privacy and Safety > Your posts > Add location information to your posts

Tagging

» More > Settings > Privacy and Safety > Audience, media and tagging > Photo tagging > Control who can tag you in photos. You can turn it off completely by using the slider or choose between "Anyone can tag you" or "Only people you follow can tag you"

Mute and block

» More > Settings > Privacy and Safety > Your X activity > Mute and block. You can block accounts, words, and notifications.

TIKTOK

Private account

» Profile > Three-line menu > Privacy > toggle Private Account.

Comments and messages

» Limit comments to friends or turn them off; restrict who can send you direct messages.

Video downloads

» Disable the option allowing others to download your videos.

Family pairing

» For teens, parents can pair accounts to manage screen time and content restrictions.

Additional Tips

» **Hard-to-find settings.** If you're following along, you'll notice that some of these settings are buried and can be challenging to find. I'm not saying that's on purpose, but you need to find and manage these settings.

» **Regularly check for changes.** These platforms frequently update their privacy interfaces, so make it a habit to revisit your settings regularly.

» **Two-Factor Authentication (2FA).** Enable 2FA wherever available to prevent unauthorized logins.

» **Phishing awareness.** Never log in via suspicious links or emails claiming to be from social networks; use only official apps or type the URL directly.

PUTTING IT ALL TOGETHER

Social media can enhance connections and creativity, but it also exposes users to scams, stalking, identity theft, and other risks. To enjoy these platforms safely:

» **Understand the risks.** From fake fundraisers to data thieves and opportunistic burglars, criminals exploit publicly available information.

» **Limit what and with whom you share.** Keep personal details off public pages; use private or selective visibility settings.

» **Remember that everything is permanent.** Even if you delete a post, it may still be accessible through screenshots or archives.

» **Customize your privacy settings.** Whether it's Facebook, Instagram, LinkedIn, or another platform, dig into the settings to ensure you share only what you're comfortable with.

» **Stay informed and vigilant.** Platforms evolve, and so do scammers. Regularly review your privacy options and adapt as needed.

With a more deliberate approach, both in what you post and how you configure your profiles, you can effectively leverage the best aspects of social media while minimizing its inherent risks. ⁑

CHAPTER TWELVE

Children's Online Safety

KIDS TODAY OFTEN USE THE INTERNET before they can even tie their shoes. The truth is that today's children will never live in a world without technology. It's here to stay. From school-issued tablets to social media apps, their lives are increasingly woven with digital tools. While the internet can be educational and entertaining, it also harbors real dangers, from oversharing personal info to encountering cyberbullying or online predators. Parents face the daunting question: How do I protect my children's well-being in a digital world?

This chapter aims to shed light on the risks children face online, examine whether the internet is truly safe for them, and explore the practical steps you can take to keep them out of harm's way. I'll also discuss monitoring tools and parental controls, both built into major platforms and available through specialized routers or third-party services.

Protecting Children Online

The web can be a magical place for learning and creativity, but it's also an adult-oriented environment where not all content is suitable for children. Young users might stumble upon explicit material, share private data, or fall victim to cyberbullying. Add to that the rise of chatrooms, social platforms, and school-issued devices, and the challenges multiply. For parents, navigating this terrain involves striking a balance between trust and freedom, on the one hand, and supervision and rule-setting, on the other.

Is the Internet a Safe Place for Kids?

The internet can be a valuable tool for kids, but it's not entirely safe without proper guidance. There are risks like inappropriate content, online predators, cyberbullying, and privacy issues. With parental supervision, safety tools, and open communication, it can be made much safer for children to use.

WHY THIS MATTERS

- » **Early adoption.** Children grow up with smartphones and tablets as part of everyday life.
- » **Unique risks.** Children may not fully grasp the permanence of online actions or the potential dangers lurking behind a digital screen.

KEY CONCERNS

Oversharing, sexting

- » **Social pressure.** Many kids feel compelled to overshare or engage in explicit communication due to peer or online influences.
- » **Privacy erosion.** Once posted, images or personal info can spread uncontrollably.

Cyberbullying

- » **Anonymity factor.** Bullies can hide behind fake profiles.
- » **Mental health impact.** Constant online harassment can be devastating, especially if children feel trapped in their digital circles.

Online predators

- » **Grooming.** Predators may pose as peers or trustworthy adults, manipulating children into dangerous situations.
- » **Chatrooms and social media.** Kids often meet strangers in interactive game lobbies or online forums.

School-issued devices

- » **Limited controls.** Even if schools install web filters, children can still circumvent them or encounter issues outside of school hours.
- » **Blurred boundaries.** Homework on a laptop can quickly pivot to social media or chatrooms.

Chatrooms

- » **Unregulated environments.** Some chat platforms have little to no moderation.

» **Identity deception.** Users can claim to be anyone, heightening risks for children.

IMMEDIATE ACTIONS

» **Open dialogue.** Encourage kids to discuss their online experiences and ask questions without fear of retribution or punishment.

» **Set boundaries.** Clarify what is acceptable to share. Refrain from sharing personal contact information, such as addresses or phone numbers.

» **Check privacy settings.** For older kids using social media, ensure their profiles are set to private and their friend lists are carefully curated.

Can You Keep Kids Safe Online?

Good news! You can help keep kids safer online by staying involved in their internet use. Parents can use parental controls, set clear rules, and talk regularly with their children about what they do online. This helps protect them from dangers like predators, cyberbullying, scams, and harmful content. Being aware and engaged is the best way to guide and protect them.

WHY THIS MATTERS

» **Empowerment vs.** restriction. An overly restrictive approach can backfire, but a lack of oversight can leave children at risk.

» **Behavioral guidance.** Teaching responsible online behavior fosters self-awareness and caution in kids.

KEY POINTS

» **Age-appropriate controls.** Young children may require stricter filtering, while teenagers may benefit from more nuanced discussions about trust and consequences.

» **Education is key.** Explaining why specific sites or behaviors are dangerous helps kids internalize rules.

» **Ongoing process.** "Set it and forget it" doesn't work. Children's interests and the online landscape are constantly evolving.

IMMEDIATE ACTIONS

> » **Regular conversations.** Ask about favorite apps or sites. Discuss potential pitfalls, such as fake profiles and inappropriate content.

> » **Lead by example.** Demonstrate good digital hygiene yourself; think before posting and regularly review your privacy settings.

> » **Adapt over time.** Reevaluate rules as children grow, offering more responsibility when they show maturity.

Are There Ways to Monitor Your Child's Online Activity?

Yes, there are ethical ways to monitor your child's online activity. You can use parental control apps, set up device screen time limits, and review browser history and app usage. The most essential tool is open communication; talk with your child about what they do online and why staying safe is so important.

WHY THIS MATTERS

> » **Visibility.** Without some awareness, you might only learn of issues after they escalate.

> » **Privacy balance.** Over-monitoring can erode trust and prompt kids to conceal their activities.

METHODS OF MONITORING

> » **Direct supervision.** Younger kids using devices in common areas while parents observe.

> » **Parental control apps.** Tools that log browsing history, limit screen time, or block certain content.

> » **Router-level monitoring.** Some routers offer usage logs that display which websites were accessed. Some also allow content category blocking. This means you can block websites that contain porn, violence, or drugs by blocking those categories and others at the router level. Remember that *no tool is 100 percent effective*, so this is not a replacement for monitoring and conversations.

KEY CONSIDERATIONS

> » **Age and maturity.** A teenager might consider constant monitoring

invasive, requiring more transparent discussions about trust and safety.

» **Legal and ethical.** Monitoring must respect your child's rights, especially as they grow older.

» **Device compatibility.** If you choose to use monitoring apps, ensure they are compatible with your child's device; check if they are iOS, Android, or Windows.

IMMEDIATE ACTIONS

» **Discuss boundaries.** If you implement monitoring, explain what you're doing and why.

» **Test the tools.** Before relying on a monitoring app, test its functionality to ensure it works seamlessly.

» **Set alerts.** Some systems send notifications if a child tries to access blocked content or chat with unknown contacts.

What Is Available to Parents for Child Internet Safety?

From network-level controls to built-in device features, parents have a range of options for safeguarding kids online.

ISP ROUTER RESTRICTIONS

» **Built-in controls.** Some internet service providers offer simple content filters or parental controls directly in the default router.

» **User and device whitelisting.** Limit which devices can connect, track usage, or block suspicious domains.

PLATFORM MANAGEMENT AND MONITORING TOOLS

<u>iOS</u>

» **Screen time.** Set daily app limits, content restrictions, and downtime schedules.

» **Family sharing.** Approve or deny app purchases and manage Apple ID settings for kids.

<u>Android</u>

» **Family link.** Manage apps, set time limits, and view essential device activity.

Windows

» **Microsoft family safety.** Allows screen time controls, activity reports, and content restrictions.

Buying a Router with More Controls

Replacing your home router with a newer model gives you better security, speed, and control. Older routers may have unpatched vulnerabilities, slower performance, and lack key features such as parental controls, guest networks, or advanced firewall settings. Aging hardware can fail or struggle with modern devices and apps. Outdated protections leave your network open to hackers, and limited functionality means you can't correctly manage or monitor your network. A new router gives you stronger defenses and more control over how your home connects to the internet.

» **Advanced parental controls.** Many modern routers and mesh systems, such as Netgear Orbi, ASUS AiProtection, and Eero, include built-in parental controls.

» **Country filtering.** Block entire regions known for malicious traffic or inappropriate sites.

» **Content and category filtering.** Pre-set categories (adult content, gambling, violence) can be blocked.

» **Malware and botnet filtering.** Some routers automatically detect suspicious connections.

» **Access control.** Restrict internet access based on device, MAC address, time of day, or user profile.

» **Third-party tools.** Services like OpenDNS (Cisco Umbrella) or Cloudflare's parental controls can add an extra layer of web filtering.

WHY THIS MATTERS

» **Easier administration.** Centralizing controls at the router can simplify management; changes apply to every device on the network.

» **Stronger filters.** Routers with built-in intelligence often update automatically to block new malicious sites.

» **Granularity.** You can set policies differently for each child's device or for different times of day.

IMMEDIATE ACTIONS

» **Evaluate existing hardware.** Check if your ISP's router or your router supports advanced parental controls.

» **Consider an upgrade.** If your router is basic, switching to one with robust filtering and logging can be worthwhile.

» **Research categories.** Decide which content categories to block or allow. For example, do you want to block all social media for younger kids or just adult sites?.

PUTTING IT ALL TOGETHER

The internet can be a fantastic resource for education, creativity, and connection, but it's not inherently safe for children. Dangers like oversharing, cyberbullying, and online predators are real and can be devastating if left unchecked. Balancing monitoring and guidance is crucial.

A combination of open communication, basic rules, and technical safeguards, including platform-specific parental controls, advanced router filters, and, at times, direct supervision, can help shield your kids from the worst the internet has to offer. You don't have to adopt an authoritarian approach; often, conversation and trust go further than any filter in building digital resilience. Yet a little strategic technology, such as a feature-rich router or family management tools, can help you enforce boundaries when the child cannot self-regulate or fully grasp the risks.

Ultimately, the goal is to empower children with the knowledge and habits to navigate the online world responsibly, knowing they have a safety net, while also learning to make informed, independent decisions. By understanding both the risks and the tools available, you can create a safer digital environment that nurtures curiosity while guarding against harm. ⁑

CHAPTER THIRTEEN

Home Office Security

THE LINE BETWEEN PERSONAL LIFE AND work can become blurry when you're juggling emails, conference calls, and company data from the comfort of your living room. This work-from-home (WFH) environment often lacks the robust security systems found in dedicated office spaces, raising the stakes for both employees and employers. Beyond the convenience, there is a real risk of data breaches, liability issues, and confusion between personal and business activities.

In this chapter, I'll explore the potential dangers of mixing work and personal use on the same devices, how to keep them separate, why it matters, and what to do if your company doesn't provide a dedicated work computer.

Challenges in Protecting Home Offices

Remote work soared worldwide, offering flexibility and saving on commute times. However, many homes aren't equipped to handle sensitive corporate data. Unsecured home networks, out-of-date personal devices, and oversharing among family members can put proprietary information at risk.

Furthermore, employees can inadvertently open malicious attachments, click phishing links, or run personal software that conflicts with corporate security policies. By establishing clear boundaries and implementing practical safeguards, you can maintain productivity while ensuring that data integrity remains intact.

Let's look at the risks and potential liabilities of conducting business on a personal device, and vice versa.

WHY THIS MATTERS

>> **Corporate data exposure.** If you store work documents on a personal computer, that machine's lesser security could open a back door

for hackers.

» **Personal privacy.** Conversely, if an employer needs to conduct security checks on a device, your private emails, photos, and personal data could be exposed to them.

» **Legal ramifications.** In regulated industries, such as finance and healthcare, mixing personal and work data may violate compliance rules, resulting in substantial fines.

» **Freedom of Information Act (FOIA).** If you work for a government agency and use your phone for work-related purposes, such as texting coworkers or supervisors about work-related issues, your device and the information on it may be included in an FOIA request and may need to be surrendered for discovery purposes.

KEY CONSIDERATIONS

» **Data breaches.** Work files stored in personal cloud accounts or local drives can get compromised if your device lacks encryption or is shared with other family members.

» **Company policy.** Some employers forbid personal usage on corporate devices or require dedicated devices for security reasons. Violating these policies could result in disciplinary action or liability.

» **Insurance and liability.** If a personal laptop used for work is stolen or compromised, you could face personal liability or insurance issues, especially if sensitive client data is involved.

IMMEDIATE ACTIONS

» **Review company policy.** Understand what your employer allows or disallows regarding device usage.

» **Check security requirements.** Are you required to use a VPN, encryption, or specific security software?

» **Keep backups.** Maintain separate backups for personal and work data to avoid overwriting or mixing essential files.

How Can You Separate the Two?

If you must use the same physical device for both personal and professional tasks, taking steps to partition your workspace is crucial.

SEPARATE ACCOUNTS, STORAGE, ACCESS

» **User accounts.** Create a dedicated user profile or account on your computer for work, using distinct login credentials separate from your home account.

» **Encrypted partitions.** On some OS (examples are Windows using BitLocker or macOS using FileVault and additional volumes), you can create separate encrypted drives or volumes for work and personal data.

» **Cloud storage.** Use a corporate-approved cloud solution — such as OneDrive for Business, Google Workspace, or corporate Box/Dropbox for work files — while storing personal photos and documents on separate personal cloud accounts.

NETWORK SEGMENTATION

» **Guest network.** Place your work devices on a dedicated Wi-Fi network separate from personal or IoT gadgets, if your router supports VLANs or multiple SSIDs.

» **VPN for work.** Ensure you use a corporate VPN for any work-related web traffic, isolating it from your browsing.

BEST PRACTICES FOR SEPARATION

» **Don't mix browsers.** Use one browser with your corporate profile and logins exclusively for work and another for personal tasks.

» **Resist "Quick Checks".** Closing your email tab before a work meeting may seem trivial, but every step counts in avoiding slip-ups. You may also save yourself from an embarrassing pop-up during an online meeting.

» **Notification settings.** Turn off personal app notifications in your "work" account environment and vice versa to limit cross-contamination.

Use Your Work Device for Work and Your Device for Your Own Business

Ideally, the simplest solution is physical separation: have one machine for work and another for personal life.

WHY THIS MATTERS

- » **Minimal overlap.** If work data never touches your device, you reduce the risk of contamination or accidental access.
- » **Clear boundaries.** It's easy to "clock out" mentally if you can close one laptop for personal time.
- » **Company security tools.** Work laptops often come preloaded with corporate security configurations, virus scanners, device management, and encryption.

WHEN THIS IS POSSIBLE

- » **Employer-provided hardware.** Many companies provide employees with laptops that are locked down to IT's standards.
- » **Personal dedication.** If you have the budget, investing in an affordable secondary work device can keep your digital life simpler and safer.

IMMEDIATE ACTIONS

- » **Request a work device.** If possible, encourage your employer to provide official hardware.
- » **Dedicated personal tools.** Use your phone or tablet for social media, streaming, or other entertainment. Avoid mixing it with enterprise communication apps unless approved by policy.
- » **Mindful practices.** If you're forced to use a single device, treat it with caution. Do not use personal software or logins on the "work account," and refrain from sharing your logins with family.

Using a Personal Device for Work

Not every organization has the resources or policies to issue separate work devices and you may need to use your personal devices. In these scenarios, you can still apply safety measures:

- » **Check for reimbursement.** Some companies might reimburse partial costs if you purchase a device that meets specific security standards. Explore grants or programs if you're in education or a smaller non-profit environment.

» **Enforce strict separation.** Set up separate OS user accounts (or even dual boot if feasible). Utilize corporate-sanctioned cloud services or remote desktop solutions to ensure that data is primarily stored on the company's server rather than on your local disk.

» **Secure everything.** Keep your OS fully updated; don't skip any patches. Employ robust anti-malware, encryption, and strong passwords or passphrases. If your employer requires specific software, such as endpoint protection or VPN, install it within the dedicated work account.

» **Communication is key.** Document your environment and the steps you take to secure it so your employer knows you're complying with best practices. Request guidelines or policies to ensure you're not operating without clear direction on critical security measures.

PUTTING IT ALL TOGETHER

Home office security is about maintaining the same vigilance you'd have in a corporate setting but within the relaxed environment of your living room or spare bedroom. Mixing personal and professional data can create a tangle of compliance risks, privacy violations, and security vulnerabilities.

Whenever possible, keep distinct devices: one for work and one for personal use. If that's not an option, at least partition your digital life with separate accounts, segmented networks, and robust security policies.

Remember:

» **Know the risks.** Liability, data leakage, and privacy issues can arise from unsegregated use.

» **Separate or segment.** Physical hardware separation is best. Otherwise, compartmentalize via OS accounts, storage locations, or network segments.

» **Stay transparent and updated.** Communicate with your employer about your home office setup, and ensure you apply all recommended security measures.

By drawing clear boundaries, even in a single-home environment, you prevent data cross-contamination, reduce your legal exposure, and protect both your employer's interests and your privacy. ⚡

CHAPTER FOURTEEN

Protecting IoT Devices

HOME AUTOMATION IS NO LONGER A futuristic concept; it's something that most of us have adopted in some form, whether it's a smart thermostat, security camera, or voice-controlled assistant. These devices can simplify everyday tasks, improve security, and increase comfort. However, as much as Internet of Things (IoT) devices can enhance your home life, they also introduce security vulnerabilities if not properly secured. Many IoT devices are poorly protected out of the box, making them easy targets for hackers if not properly secured.

In this chapter, I'll walk through the essential steps to protect your IoT devices from the threats they may face. These devices often communicate over your home Wi-Fi, and once compromised, they could expose your data and network or even grant unauthorized access to your home. The good news is that protecting them doesn't require a tech degree — just some basic adjustments that will go a long way in improving your digital security.

Why Securing IoT Devices Is Crucial

When you buy a smart device such as a camera, doorbell, fridge, or even a toothbrush, it often comes with default settings that prioritize ease of use over security (remember the SUC triangle?). These factory settings are convenient, but they also make it easier for malicious actors to hack into your devices and your home network.

» **Increased attack surface.** The more devices connected to the internet, the more entry points for potential attacks. A single unprotected device can be a backdoor for criminals.

» **Sensitive data.** Many IoT devices collect sensitive personal data, including your location, activity patterns, and conversations. Protecting these devices prevents unauthorized access to that information.

» **Botnets.** Hackers can control poorly secured IoT devices and use them as part of a botnet, a network of compromised devices used to launch attacks on other systems.

Change Default Passwords Immediately

One of the easiest ways for hackers to access your IoT devices is by exploiting the factory-set passwords that come with many of these devices. By simply changing the default password, you can drastically reduce the likelihood of unauthorized access.

WHY THIS MATTERS

Default passwords are often widely known or easily guessed. Cybercriminals frequently utilize databases of default credentials to attempt unauthorized logins to IoT devices.

STEP-BY-STEP GUIDE

» **Log in to your device settings.** Refer to the user manual or check the device's setup guide to find the IP address or app interface for accessing the device's settings. For example, many smart cameras or thermostats have a mobile app you use to configure settings.

» **Change the default username and password.** Find the login section in the settings. The default username is often something like "admin" or "user," while the default password may be a simple one, such as "1234" or the model number. Change both the username and password to something more complex and unique. Think password manager.

» **Use a strong password.** Create a password that combines uppercase and lowercase letters, numbers, and symbols. Avoid using easily guessable information, such as your name, birthday, or simple patterns.

» **Document and store your new passwords.** Use a password manager to securely store the new credentials.

BEST PRACTICES

> » **Unique credentials for every device.** Avoid reusing passwords across different devices or services.
>
> » **Change periodically.** Set a reminder to change your device passwords every 6 to 12 months. This simple but effective step is critical to securing your IoT devices against potential attacks.

Update Firmware Regularly

Firmware is the software built into your IoT devices that controls their functionality. Manufacturers frequently update firmware to patch security vulnerabilities, fix bugs, and improve performance. If you don't update your devices, they may be vulnerable to new security flaws.

WHY THIS MATTERS

Cybercriminals actively search for vulnerabilities in outdated devices. If a device no longer receives security patches, it becomes a prime target for exploitation. If a device is no longer supported, plan for its replacement.

STEP-BY-STEP GUIDE

> » **Check for firmware updates.** Most IoT devices can check for updates either automatically or manually. If your device has an app or web portal, look for a "Firmware Update" or "Software Update" option in the settings menu. Alternatively, visit the manufacturer's website to see if updates are available.
>
> » **Enable automatic updates (if available).** Many modern devices allow you to enable automatic updates so you don't have to check for patches manually. This ensures you're always running the latest version.
>
> » **Manually install updates.** If your device doesn't support auto-updating, check for updates regularly — every 1 to 3 months — to ensure it is patched with the latest security fixes.
>
> » **Reboot devices after updates.** Some firmware updates may require a device reboot to take effect. Be sure to restart the device after installing updates.

BEST PRACTICES

> » **Keep track of updates.** Set a calendar reminder to check for updates at least every quarter.

> » **Don't skip firmware updates.** Just as you would with software updates on your phone or computer, treat IoT firmware updates as essential to maintain security.

Disable Unnecessary Features

Many IoT devices have features that may be convenient but aren't always necessary. If left enabled, these extra features can introduce additional security risks, as they provide more opportunities for an attacker to gain control.

WHY THIS MATTERS

The more active features a device has, the more potential vulnerabilities it may carry. Disabling any features you don't actively use reduces the attack surface, which will make it harder for attackers to exploit weaknesses.

STEP-BY-STEP GUIDE

> » **Identify features you don't need.** Review the settings of each IoT device to determine which features are essential for your usage. Standard features to disable include remote access, voice assistants, and Bluetooth if you're not using these services.

> » **Turn off remote access.** If you don't need to control your device remotely, disable features like "remote access" or "cloud connection." These settings are often enabled by default. Certain devices like alarm base stations may require remote access for them to function correctly during an emergency. Refer to your device's manuals and documentation for specific details.

> » **Disable voice assistants or Bluetooth.** Many smart speakers or home assistants have built-in voice features or Bluetooth connectivity. If you don't use these options regularly, it's wise to turn them off when not in use to limit exposure.

> » **Check your privacy settings.** Many IoT devices collect data about

your habits or even record audio or video. Go through privacy settings to ensure you only share the information you're comfortable with.

BEST PRACTICES

» **Use local control.** If your device offers this option, it is recommended to control it locally (on your home network) rather than remotely via the internet.

» **Turn off when not in use.** If a device has a feature you use infrequently, consider turning it off when not in use, especially when you're away from home.

Disabling unnecessary features enhances the overall performance and lifespan of your devices by reducing the amount of functionality they need to maintain, thereby improving security.

Use a Separate Wi-Fi Network for IoT Devices

As discussed in Chapter Three, "Securing Your Home Network," one of the most important steps you can take to secure your home network is to create a guest network for less trusted devices. IoT devices, such as smart TVs, security cameras, and doorbells, often lack robust security, and isolating them from your primary network helps prevent attackers from gaining access to your most sensitive devices.

By segmenting your IoT devices, you reduce the overall risk to your network and ensure that a compromised device doesn't jeopardize your entire home system.

WHY THIS MATTERS

IoT devices are often less secure than your primary devices, such as computers and smartphones. Keeping them on a separate network helps contain potential risks. If a hacker compromises a smart camera or thermostat, they won't have access to your primary network, where your devices and files reside.

STEP-BY-STEP GUIDE

» **Set up a guest network.** If your router supports multiple SSIDs, create a secondary network exclusively for your IoT devices. Look for an option like "Guest Wi-Fi" or "Network Segmentation" in your router's settings.

» **Unique name.** Name the network something distinct, such as "IoT_ Network" or "HomeDevices."

» **Secure the guest network.** Set a strong password for the guest network, similar to the one used for your primary network, and ensure that WPA2 or WPA3 encryption is enabled.

» **Connect IoT devices to the separate network.** When setting up new IoT devices, connect them to the newly created guest network rather than your primary Wi-Fi.

» **Monitor devices on both networks.** Review connected devices on both networks regularly to ensure that no unauthorized devices are using your guest or primary network.

BEST PRACTICES

» **Isolation is key.** Many routers allow you to isolate traffic on the guest network, which means devices can't communicate with your primary network. If available, enable this feature.

» **Use Virtual Local Area Networks (VLANs).** For advanced users, setting up VLANs on your router can provide even more robust isolation between devices.

PUTTING IT ALL TOGETHER

IoT devices are powerful tools that make life more convenient, but they can also expose users to potential security vulnerabilities without proper protection. By taking a few simple steps, changing default passwords, updating firmware, disabling unnecessary features, and setting up a separate network, you can significantly reduce the risks posed by these devices.

Securing IoT devices may require some initial effort, but once these steps are taken, your devices will be significantly safer, and the overall integrity of your home network will be enhanced. By following this approach, you can enjoy the benefits of connected living without compromising your security. ⚌

CHAPTER FIFTEEN

Smartphone Security

SMARTPHONES ARE OFTEN OUR MOST FREQUENTLY used devices, handling everything from banking to personal communication. Securing them is crucial.

Smartphones have become an indispensable part of our daily lives. From managing finances and personal communication to storing sensitive information such as passwords, photos, and health data, our phones have become the central hub of our digital existence. This convenience, however, comes with its own set of risks. Smartphones can be vulnerable to unauthorized access, malware, and digital theft if not properly secured. Whether you use an Android phone, iPhone, or other smart devices, it's essential to implement basic security measures to protect your data.

This chapter will discuss how to secure your smartphone against potential threats. From simple steps like locking your screen to more advanced settings like managing NFC or using a VPN on public Wi-Fi, I'll cover things that help safeguard your device.

The Cost of Smartphone Insecurity

Smartphones store an immense amount of sensitive data, including banking apps, private emails, photos, contacts, and location information. A compromised phone could lead to identity theft, financial loss, or a complete invasion of your privacy. Furthermore, smartphones are constantly connected to the internet and various networks, making them particularly vulnerable to malicious attacks. It's not just about preventing someone from stealing your phone; it's about preventing unauthorized access to the digital footprint that lives within it.

LOCK YOUR SCREEN

The first and most essential step in securing your smartphone is to ensure that it's locked with a PIN, passphrase, or biometric lock. This prevents unauthorized access to your phone in case it's lost or stolen. A locked screen is your first line of defense; don't make it easy for someone to access your phone.

WHY THIS MATTERS

Without a lock screen, anyone who picks up your phone can access all your personal information immediately. A simple screen lock can prevent this from happening and is a strong first line of defense.

STEP-BY-STEP GUIDE

» **Choose a strong screen lock.** On Android, go to Settings > Security and Privacy > Device unlock. On iPhone, go to Settings > Face ID and Passcode or Touch ID and Passcode.

» **Enable biometrics (optional but recommended).** Most modern phones support biometric authentication, including facial recognition and fingerprint scanning. Enable this feature for quicker and more secure access to your device. Register multiple fingerprints in case you get a cut or are wearing a cast.

» **Avoid simple passcodes.** Never use simple passcodes, such as "1234" or your birthday. Instead, opt for longer passcodes or biometric options.

BEST PRACTICES

» **Enable lock on timeout.** Set your phone to automatically lock after a short period of inactivity, ensuring it remains secure even if you forget to lock it manually.

» **Biometric backup.** Even with biometric access, always have a strong PIN or passcode as a backup in case your biometric data can't be read.

Bluetooth and Sharing

Bluetooth is a powerful feature, but it can pose a security risk when left open. Many devices, including phones, can automatically connect to nearby Bluetooth-enabled devices, which may expose your data or leave your phone open to attack. Keeping Bluetooth off when not in use and limiting sharing capabilities ensures your phone isn't exposed to unwanted or malicious connections.

WHY THIS MATTERS

Hackers can exploit unsecured Bluetooth connections to transfer malicious files or access your data. Keeping Bluetooth off when not in use significantly reduces these risks.

STEP-BY-STEP GUIDE

» **Keep Bluetooth off when not in use.**
  - » Android: Swipe down from the top of the screen and turn off Bluetooth.
  - » iPhone: Open the Control Center by swiping down from the top right.
  - » Tap on the Bluetooth / AirPlay / Wi-Fi widget to expand it.
  - » Tap directly on the blue colored Bluetooth icon. Tapping on the Bluetooth widget opens the device selector.
  - » Tap directly on the Bluetooth icon to turn Bluetooth ON or OFF.

» **Be cautious when sharing files.** Avoid sending or receiving files via Bluetooth unless you trust the device and connection.

» **Use Bluetooth pairing mode.** Set Bluetooth to pairing mode when connecting to new devices so that only devices you approve can connect.

BEST PRACTICES

» **Visibility settings.** Set your device to non-discoverable mode when not using Bluetooth, preventing other devices from detecting it.

» **Review paired devices.** Periodically check your list of paired devices and remove anything you don't recognize.

NFC (Near Field Communication)

NFC enables smartphones to communicate wirelessly over short distances, allowing for mobile payments and tap-to-share features. While useful, NFC can also be a security risk if enabled unnecessarily.

WHY THIS MATTERS

With NFC enabled, other devices can scan your phone, potentially exposing your personal information or enabling unauthorized transactions.

STEP-BY-STEP GUIDE

» Disable NFC when not in use:
  » Android: Go to Settings > Connections > NFC and Payment and toggle it off.
  » iPhone: Go to Settings > Wallet and Apple Pay and toggle NFC off.
» Use NFC only for trusted services:
  » Only enable NFC when needed, such as when making payments or transferring data to trusted devices.
  » Lock your phone when using NFC
  » If you must use NFC, ensure your phone is locked with a secure method (such as a PIN or fingerprint) to protect against unauthorized transactions.

BEST PRACTICES

» **Pay attention to payment apps.** If you are using mobile payment systems, ensure the app is configured securely and verify the device's authenticity before tapping.
» **Disable.** Disabling NFC when not needed and securing it during use will prevent unauthorized parties from accessing your device or initiating unwanted transactions.

Joining Public Wi-Fi Networks

Public Wi-Fi networks, such as those in coffee shops, libraries, or airports, are convenient but often insecure. Connecting to these networks without caution can expose your phone to security risks. Using a VPN and disabling auto-connect options can help mitigate the risks associated with connecting to untrusted Wi-Fi networks.

WHY THIS MATTERS

Public Wi-Fi networks may be unencrypted, which means that data sent over them can be intercepted by anyone else on the same network. Hackers can use this to access sensitive information, including passwords and financial details.

STEP-BY-STEP GUIDE

- » **Avoid connecting to unsecured networks.** If a Wi-Fi network does not require a password or is labeled "open," avoid connecting.
- » **Use a VPN on public Wi-Fi.** If you must use public Wi-Fi, consider using a trusted VPN (Virtual Private Network) to encrypt your internet connection. A VPN helps protect your data from being intercepted.
- » **Disable auto-connect to Wi-Fi networks.**
  - » Android: The settings for this will vary based on the phone manufacturer and the version of Android installed.
  - » iPhone: Go to Settings > Wi-Fi and toggle off Auto-Join for unknown networks.

BEST PRACTICES

- » **Enable HTTPS.** When browsing the internet on public Wi-Fi, ensure that the websites you visit use HTTPS (indicated by a padlock icon), which indicates a secure connection.
- » **Use trusted networks.** Whenever possible, use your mobile data or a trusted, encrypted Wi-Fi network instead of public networks.

App Downloads and Updates

Keeping your apps up to date is essential for securing your smartphone. Apps can contain vulnerabilities patched through updates, so neglecting updates can expose your phone to potential threats.

WHY THIS MATTERS

Like operating systems, apps receive updates that fix security vulnerabilities, add new features, and enhance performance. Keeping apps up to date helps protect your phone against malware and other threats.

STEP-BY-STEP GUIDE

- » **Only download from trusted sources.** Ensure you only download apps from official app stores, as mentioned earlier in this book.
- » **Enable auto-updates for apps.**

> » Android: Go to Google Play Store > Settings > Network preferences > Auto-update apps and select your preferred update option.
>
> » iPhone: Go to Settings > App Store and enable App Updates under Automatic Downloads.

» **Manually check for app updates.** Occasionally, manually check for updates in your app store to ensure all apps are up to date and your settings are accurate.

BEST PRACTICES

> » **Review permissions after updates.** Some app updates may change permissions or add new features. After updating, review the permissions each app has access to.
>
> » **Stay current.** Keeping apps updated helps protect your phone against the latest threats and ensures that the apps you use remain secure.

Regular Security Audits

Regular phone security audits ensure that your apps and settings remain secure over time. This involves reviewing the apps you've installed and the permissions they've been granted and removing anything you no longer use or need. Performing regular security audits ensures that your phone stays protected as new threats emerge and old vulnerabilities are patched.

WHY THIS MATTERS

Over time, apps can accumulate permissions or data that you no longer need or want. Regularly auditing your device ensures you're not inadvertently giving away more access than necessary.

STEP-BY-STEP GUIDE

> » **Review installed apps.** Go to your device's App Settings to see a list of all installed apps. Remove any that you no longer use.
>
> » **Review app permissions.** As mentioned earlier, regularly check app permissions to ensure that apps don't have unnecessary access to your data or features.
>
> » **Keep your device clean.** Uninstall unused apps, clear browser caches,

and remove any sensitive data you no longer need.

BEST PRACTICES

> » **Install only what you need.** Avoid cluttering your phone with apps you rarely use, as this increases potential attack vectors.
> » **Clear your cache.** Regularly clear your app caches and browser data to prevent excess data from being stored on your device.

PUTTING IT ALL TOGETHER

Smartphone security isn't just about locking your screen; it's about implementing a series of thoughtful, proactive steps to secure your device against unauthorized access and data breaches. By following the practices outlined in this chapter locking your screen, managing Bluetooth and NFC settings, avoiding unsecured Wi-Fi, regularly updating apps, and conducting regular security audits you can help ensure your smartphone remains safe from threats.

Your phone is a powerful tool, and securing it is an ongoing process. With the steps provided here, you'll be better equipped to maintain a secure mobile environment and protect the sensitive data stored on your device. Stay vigilant and remember that securing your smartphone is key to keeping your digital life safe. ⚡

CHAPTER SIXTEEN

Cross-Platform Security

EACH PLATFORM, SUCH AS APPLE, MICROSOFT, Google, and so on, has different security features; learn to utilize them effectively.

In today's multi-device world, many individuals and families use a combination of different operating systems (OS): Apple's macOS and iOS, Microsoft's Windows, and Google's Android. Each platform has its own set of security tools and features that help protect your data and privacy. While these platforms differ in many ways, they all offer robust security features that are built into the system. The key to staying secure is learning how to use these tools properly, regardless of which platform you're on.

In this chapter, I'll explore how to secure your devices across different platforms, focusing on the built-in security tools provided by Apple, Microsoft, and Google. I'll cover how to enable these tools, manage passwords securely, and track devices in case they get lost or stolen.

The Importance of Security on Varying Devices

With the rise of multiple devices, ranging from desktops and laptops to smartphones and tablets, security has become more complex than ever. It's no longer enough to focus on securing one device or one operating system. If you use Apple, Microsoft, and Google devices in your daily life, it's essential to understand how to keep each platform secure.

» **Platform-specific threats.** Each platform may have unique vulnerabilities. Ensuring that security tools are correctly configured can help mitigate these risks.

» **Data syncing across devices.** Many people use a combination of devices to access email, bank accounts, social media, and other services. Consistent security practices ensure that your data stays protected across platforms.

» **Device loss.** Losing a device is a significant risk. Many platforms have built-in tools to help you locate or remotely erase your device, preventing unauthorized access.

By leveraging the security tools built into Apple, Microsoft, and Google platforms, you can create a cohesive and secure digital environment across all your devices.

Enable Built-in Security Tools

Each platform offers built-in security tools, including firewalls, malware scanners, and encryption tools, to help protect your data from various attacks. Ensuring these tools are enabled is a crucial step in securing your devices.

WHY THIS MATTERS

Built-in security features are designed to protect you from common threats such as viruses, malware, and unauthorized access. If these tools aren't activated or properly configured, your devices may be vulnerable.

STEP-BY-STEP GUIDE

Apple macOS:

» **Ensure the firewall is enabled.** Go to System Settings > Network > Firewall, then click Turn On Firewall.

» **Use FileVault.** Enable FileVault for full disk encryption: Go to System Settings > Privacy and Security > FileVault, then click Turn On FileVault.

Apple iOS:

» **Use Find My iPhone.** This enables you to locate or lock your device remotely: Go to Settings > [Your Name] > Find My, and toggle Find My on.

» **Enable iCloud Backup for secure backups.** Go to Settings > [Your Name] > iCloud > iCloud Backup.

Windows

» **Windows Defender.** Ensure that Windows Defender Antivirus is enabled: Go to Settings > Privacy and Security> Windows Security > Virus and Threat Protection.

» **Firewall.** Turn on the Windows Firewall: Go to Settings > Privacy and security > Windows Security > Firewall and network protection. Then click Turn Windows Firewall on or off.

» **BitLocker.** Use BitLocker for disk encryption. Go to Settings > Privacy and Security > Windows Security > Device Security > Data Encryption.

Google (Android)

» **Google Play Protect.** Enable Google Play Protect for malware scanning. Go to Settings > Google > Security > Google Play Protect, then select "Scan apps with Play Protect."

» **Find My Device.** Enable Find My Device to locate or erase your phone remotely: Go to Settings > Security > Find My Device and toggle it on.

BEST PRACTICES

» **Enable all available security features.** Don't disable built-in security tools, even if they seem unnecessary. The exception would be if they are conflicting with other third-party security tools you have installed, and you are troubleshooting the problem.

» **Update software regularly.** Ensure all your software is up to date to keep your built-in security tools current.

» **Device protection.** Enabling these built-in tools helps protect your devices from unauthorized access, viruses, and malware, reducing the risk of a breach.

Sync Passwords Securely

Passwords are the key to your digital life, and syncing them securely across devices is crucial. Fortunately, Apple, Microsoft, and Google all offer password management tools that enable you to store and sync passwords across multiple platforms securely. By securely syncing passwords across your devices and using trusted password managers, you can protect your credentials from unauthorized access.

WHY THIS MATTERS

Storing passwords in an unsecured manner, such as in a text file, email, or browser, can put your accounts at risk if someone gains access to your device. Secure password management ensures that your credentials are stored in an encrypted and safe environment.

STEP-BY-STEP GUIDE

Apple (iCloud Keychain)

» **Enable iCloud Keychain.** This will securely store and sync your passwords across all your Apple devices: Go to Settings > iCloud > Passwords, then toggle on Sync.

Microsoft

» **Microsoft Edge.** This browser features a built-in password manager that securely stores passwords. Ensure that syncing is enabled: Go to Settings > Profiles > Sync > Passwords, then turn on Password Sync.

» **Windows Credential Manager.** Store and manage your Windows passwords in Control Panel > Credential Manager.

Google

» **Google Password Manager.** Store passwords in your Google account and sync them across your devices. In Chrome, navigate to the three-dot menu > Passwords and Autofill> Google Password Manager, and then toggle "Offer to save passwords and passkeys."

BEST PRACTICES

» **Use a third-party password manager.** However, if the expense or complexity doesn't make that a viable option, consider using these tools.

» **Use strong, unique passwords.** Use a mix of letters, numbers, and symbols for each password. Avoid reusing passwords across accounts.

» **Enable Two-Factor Authentication.** In addition to using secure password managers, enable two-factor authentication (2FA) on accounts that support it for an added layer of protection.

Enable Device Location Tracking

Device location tracking is an essential tool for ensuring that you can locate your devices if they are lost or stolen. Apple, Microsoft, and Google all offer location tracking features that enable you to find, lock, or remotely wipe your device to protect your data.

WHY THIS MATTERS

If your phone, tablet, or laptop is lost or stolen, you can use location tracking to find it, lock it, and protect your data. Enabling this feature ensures that your devices are recoverable and your information remains safe.

STEP-BY-STEP GUIDE

Apple (Find My iPhone)

» **Enable Find My iPhone.** Go to Settings > [Your Name] > Find My > Find My iPhone and toggle it on.

» **Locate, lock or erase.** Use Find My to locate, lock, or remotely erase your device through the Find My app or iCloud.com.

Microsoft (Find My Device)

» **Enable Find My Device.** Go to Settings > Privacy and security > Location and toggle on Location services. Then, go to Settings > Privacy and Security > Find My Device to enable tracking.

Google Android (Find My Device)

» **Enable Find My Device.** Go to Settings > Security and privacy > Device finders > Find My Device and toggle it on.

» **Locate, lock or erase.** Use the Find My Device app or visit google.com/android/find to locate, lock, or erase your device remotely.

BEST PRACTICES

» **Enable location tracking for all devices.** Don't just limit location tracking to your phone; enable it on your tablets, laptops, and other devices as well.

» **Set up remote wipe.** Enable the remote wipe feature for your devices to ensure your data is erased if your device is lost or stolen.

» **Track and protect.** By enabling device location tracking, you can track your devices if they are lost or wipe and protect your data from theft.

PUTTING IT ALL TOGETHER

Cross-platform security is crucial for individuals who use multiple devices with varying operating systems. Apple, Microsoft, and Google all provide robust security features that help protect your data across platforms. By enabling built-in security tools, syncing passwords securely, and utilizing device location tracking, you can establish a robust defense against unauthorized access and data theft, regardless of the platform you're using.

By learning how to utilize the security features provided by each platform and implementing best practices, you can ensure that your devices remain secure, your data is protected, and you're always in control of your digital life. ⁑

CHAPTER SEVENTEEN

Automotive Software Security

MODERN VEHICLES ARE CONNECTED TO THE internet, making them potential targets. Today's vehicles are no longer just machines for getting from point A to point B; they are increasingly sophisticated, connected, and driven by software. Features like GPS navigation, advanced driver assistance systems (ADAS), in-car entertainment, and even remote vehicle management apps are now commonplace. While these technologies add immense convenience and improve driving experiences, they also introduce new vulnerabilities. A connected vehicle is a potential target for hackers who could exploit security flaws in the vehicle's software or its connections to the broader internet.

In this chapter, I will discuss essential practices for securing the software in your vehicle, including keeping it up to date, managing third-party apps, and disabling unused connectivity features. By following these steps, you can ensure that your car's digital systems remain as secure as possible.

Staying Ahead in Automotive Software

Modern vehicles come equipped with a wide range of software-driven systems that improve performance, safety, and convenience. However, like any device connected to the internet, they are vulnerable to attacks if not properly secured. Some of the potential risks include:

» **Remote access exploits.** Many vehicles now offer remote access features, such as remote start, tracking, and diagnostics. If compromised, these systems can be used to hijack a car or steal sensitive information.

» **In-vehicle network compromise.** Vehicles feature complex internal networks that manage critical systems such as braking, steering, and powertrain. If these systems are hacked, it could put your safety at risk.

» **Third-party apps.** Many vehicles allow third-party apps to connect to the vehicle's software. These apps may contain vulnerabilities that could be exploited to gain access to the car's systems.

Ensuring your vehicle's software is secure is an essential part of maintaining your security and safety on the road.

Keep Vehicle Software Updated

Just like smartphones and computers, modern vehicles require regular software updates to maintain security. These updates often patch vulnerabilities, improve functionality, and add new features.

WHY THIS MATTERS

Outdated vehicle software can contain security holes that hackers can exploit to gain access to the car's systems. Just like any other software, vehicle software needs to be updated regularly to protect against new threats.

STEP-BY-STEP GUIDE

» **Check for software updates.** Most modern vehicles allow you to check for software updates directly through the vehicle's settings menu. Alternatively, you can check via the manufacturer's mobile app or website. For cars with in-vehicle connectivity, check the infotainment system settings for firmware updates.

» **Schedule updates.** Many manufacturers will notify you when an update is available. Ensure that you install updates as soon as they are released, particularly those that address security vulnerabilities.

» **In-dealership software updates.** Some updates may require a visit to your car dealership or service center. During service visits, ask your technician to verify that your vehicle's software is up to date, especially if it is equipped with advanced features such as autonomous driving or remote access.

BEST PRACTICES

» **Enable automatic updates.** If your vehicle supports automatic updates, enable this feature to ensure that updates are applied without delay.

» **Monitor recall notices.** In addition to software updates, be aware of recalls related to software vulnerabilities.

» **Protect you and your vehicle.** Regular software updates ensure that your car's digital systems remain secure and are functioning optimally, protecting both you and your vehicle.

Be Cautious with Third-Party Car Apps

Third-party apps, such as those used for vehicle diagnostics, remote control, or in-car entertainment, offer additional features but can also introduce security risks. Apps that connect to your vehicle's systems could potentially serve as an entry point for attackers. Being cautious with third-party apps ensures that you're not exposing your vehicle to unnecessary risks and that any apps you use are trustworthy.

WHY THIS MATTERS

Third-party apps may not have the same security standards as those developed by the vehicle manufacturer. These apps could have vulnerabilities that cybercriminals can exploit, allowing unauthorized access to your vehicle's digital systems.

STEP-BY-STEP GUIDE

» **Use official apps from the manufacturer.** Stick to apps provided directly by your vehicle's manufacturer whenever possible. These apps are typically more secure because they are designed specifically for your vehicle and receive regular updates.

» **Research third-party apps.** If you decide to use third-party apps, conduct thorough research to ensure they are trustworthy and reliable. Read reviews, check for updates, and verify that the app comes from a reputable source.

» **Limit permissions.** Review the permissions each app requests. For instance, if an app asks for access to vehicle control features but doesn't require it for basic functionality, consider denying that permission.

BEST PRACTICES

» **Avoid unnecessary apps.** If you don't need a third-party app, don't install it. Fewer apps mean fewer potential vulnerabilities.

» **Be wary of free apps.** Free apps are often monetized by collecting user data, which may compromise your privacy or security.

Disable Unused Connectivity Features

Many modern vehicles come equipped with features such as remote access, Wi-Fi hotspots, and Bluetooth connectivity. While these features are convenient, they also increase your vehicle's attack surface. If you don't need these features, it's a good idea to disable them.

WHY THIS MATTERS

Unused connectivity features, such as remote vehicle access or Wi-Fi hotspots, can act as entry points for attackers. By disabling features you don't use, you reduce the number of potential vulnerabilities that can be exploited.

STEP-BY-STEP GUIDE

» **Turn off remote access.** If you don't use remote access features like remote start or remote vehicle tracking, disable them in your vehicle's settings or via the manufacturer's mobile app.

» **Disable Wi-Fi hotspot.** If your car is equipped with a Wi-Fi hotspot, turn it off when not in use. This prevents hackers from accessing your car's internal network via an unsecured Wi-Fi connection.

» **Turn off Bluetooth when not in use.** If your vehicle has Bluetooth connectivity, ensure it is disabled when not in use. This will prevent unauthorized devices from pairing with your car.

BEST PRACTICES

» **Use a strong password for Wi-Fi.** If you need to use your vehicle's Wi-Fi hotspot, ensure that it is secured with a strong password.

» **Revisit connectivity features periodically.** Over time, you may forget to turn off connectivity features you no longer use. Regularly revisit your vehicle's settings to ensure that only necessary features are active.

» **Reduce vulnerabilities.** Disabling unused connectivity features helps reduce your vehicle's vulnerability to potential attacks and ensures that only the features you use are active.

PUTTING IT ALL TOGETHER

Automotive-based software security is just as crucial as securing other devices in your life. By keeping your vehicle's software up to date, exercising caution with third-party apps, and disabling unnecessary connectivity features, you can help protect your vehicle's digital systems from malicious threats. Modern cars are equipped with cutting-edge technology that enhances your driving experience, but just like any connected device, they require proactive security measures.

By following these steps, you can ensure that your vehicle's software remains secure, reducing the risk of unauthorized access or data theft. In an increasingly connected world, securing your vehicle is a crucial part of maintaining your overall digital security. ⁑

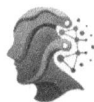

CHAPTER EIGHTEEN

Connecting the Dots

CONGRATULATIONS, YOU'VE REACHED THE FINAL CHAPTER, where I bring together all the parts of technological self-defense. By now, you've explored everything from managing passwords and browsing securely to segmenting your home network and staying safe on social media. If you feel a bit overwhelmed, remember it's not about perfection or paranoia but continuous awareness and ongoing adjustments.

To sum things up, I'll highlight the key takeaways, emphasize why layered security *(did you think I forgot about this?)* remains the best strategy, and discuss how to keep your knowledge fresh in an evolving cybersecurity landscape. I'll also encourage you to share what you've learned with family, friends, and colleagues because security is truly a team effort.

Key Takeaways From This Book

After reading through the various chapters, you see that fundamental lessons are repeated:

» **Layered security.** No single measure — such as antivirus software alone — can ensure 100 percent safety. Combining strong passwords, multi-factor authentication, encryption, and awareness significantly enhances your defense.

» **Segmentation and separation.** Whether it's separate user accounts on devices or VLANs for IoT gadgets, isolating different parts of your digital life reduces your overall risk.

» **Vigilance over fear.** You don't need to unplug from the internet. You need to stay aware that scams, breaches, and threats are constantly evolving, so your vigilance should, too.

» **Minimize oversharing.** Whether on social media or with unsolicited emails, be mindful that any info you put online can be used for malicious purposes.

» **Regular maintenance.** Simple habits, such as updating your apps, operating systems, and router firmware, can close vulnerabilities before criminals exploit them.

Living in Constant Fear Isn't the Goal

I've covered serious threats, identity theft, phishing, and data breaches. Yes, these are very serious topics, but this is not about being paranoid at every turn. It's about being aware and being prepared. Instead of panicking, use:

» **Realistic caution.** Recognize that threats exist, but adopt calm, informed strategies, such as verifying suspicious links, creating strong passwords, and using encryption.

» **Balanced approach.** Security measures shouldn't disrupt your daily life. Strive for a comfortable balance that aligns with your personal risk tolerance.

» **Empowerment.** Knowledge reduces anxiety. When you know how to spot scams or strengthen a network, you'll feel in control.

» **Cross-referencing.** Don't be afraid to revisit earlier chapters as you implement new layers. For instance, if you're setting up a VLAN, confirm that you've already secured your router.

» **Ongoing evolution.** Treat this as a living reference. Threats change, so periodically skim relevant sections (such as phishing or identity theft) to refresh your memory and adapt to new scams or vulnerabilities.

It Really Is About Layered Security!

Layered security refers to the practice of stacking multiple defensive measures, ensuring that if one fails, the others remain intact. For example:

» **Strong passwords + 2FA + App/OS Updates + Awareness.** Even if your password is stolen, 2FA or updated system patches can stop an attacker.

» **Encrypted storage + physical security (lock your devices, shred documents). Even if someone gains physical access to your hardware, encryption and locked cabinets can protect sensitive data.

» **Network segmentation, threat monitoring, and safe browsing.** Even if one IoT device is hacked, segmentation can confine the problem to that specific VLAN. In contrast, threat monitoring enables you to identify and address potential intrusions quickly.

Strategies to Stay Updated

Cyber threats evolve at breakneck speed and you must stay as current as possible. What you learn today may be irrelevant tomorrow. Attackers continually develop new malware, phishing techniques, and methods to exploit emerging technologies, such as AI or deepfakes. Here are a few pointers to stay updated:

» **Tech news and newsletters.** Follow reputable cybersecurity blogs, podcasts, or newsletters. You might want to check out Krebs on Security and SANS NewsBites.

» **Professional communities.** If you work in IT or a related field, professional groups and forums can highlight new vulnerabilities or best practices.

» **Continuing education.** Attend local workshops, watch online webinars, or take short courses on security topics.

» **Device alerts.** Enable automatic updates or at least set reminders to check for patches regularly.

Keeping Your Knowledge Current

» **Set Google alerts.** For terms like "data breach," "phishing scam," or brand names you trust, so you're notified of significant incidents when they happen.

» **Follow official channels.** Governments and major software vendors, such as Microsoft, Apple, and Google, often publish security bulletins.

» **Practice hands-on.** If you're a technical person, experiment with different security tools, password managers, or home firewall settings in a safe environment. Everyone learns and remembers differently. Don't be afraid to supplement this book with some hands-on training.

Share Your Knowledge with Others

» **Security is a community effort.** By spreading awareness, you not only elevate your defenses but also protect family, friends, and coworkers who might otherwise be vulnerable targets.

» **Family check-ins.** Assist older relatives or less tech-savvy friends in identifying phishing scams, setting up password managers, and performing essential device maintenance.

» **Workplace awareness.** If your office doesn't have a robust cybersecurity program, consider volunteering to share tips, such as safe email practices or how to identify suspicious links.

» **Parent and child communities.** For those with children, help other parents navigate safe online practices and device usage guidelines.

PUTTING IT ALL TOGETHER

As you close this book, remember: technology evolves fast. The threats you prepared for last year might morph into new forms tomorrow.

The good news? Most attackers bank on ignorance and complacency. By staying curious, applying layered security, and regularly updating both your tools and knowledge, you tilt the odds firmly in your favor.

KEY POINTS

» **Security is a journey, not a one-time fix.** Periodically review your practices, run scans, and adjust settings as needed.

» **Balance is key.** Don't let fear drive you offline; let awareness guide you toward safer digital habits.

» **Pay it forward.** Share what you've learned to create safer online communities for everyone.

With the right mindset and a bit of diligence, you can enjoy the vast benefits of modern technology while minimizing risks. After all, cybersecurity isn't about hiding; it's about living confidently in a digital world that you've made just a bit more secure. ‡

CHAPTER NINETEEN

Ongoing Security Practices

By this point, you've learned that cybersecurity is an ever-evolving field, and keeping your home network, devices, and online accounts secure demands continuous effort. It's not about setting up a few safeguards once and then forgetting about them. Hackers and cybercriminals are constantly devising new tactics, and technology changes rapidly as well. The best way to defend against emerging threats is to stay informed, periodically review your security measures, and maintain a proactive approach to safeguarding your digital life.

In this final chapter, I'll outline the ongoing practices that will help you stay safe over the long haul. You'll learn how to monitor new threats, reinforce good security habits, and ensure that everyone in your household knows how to identify and avoid common online hazards.

Stay Informed About New Threats

One of the biggest challenges in cybersecurity is that the landscape is constantly evolving. New malware, phishing schemes, and vulnerabilities emerge regularly, so it's essential to stay up to date. It's no exaggeration to say that cybersecurity threats and defenses change daily. Not all of them, but enough.

WHY THIS MATTERS

Even the most robust security measures can be compromised by emerging types of attacks. Cybercriminals adapt quickly; staying informed helps you adapt just as fast.

STEP-BY-STEP GUIDE

> » **Follow cybersecurity news.** Read reputable technology blogs or news

outlets that cover the latest breaches and scams. Examples include mainstream tech publications and dedicated security websites.

» **Subscribe to alerts.** Some operating systems and security software offer alert systems that notify you of significant threats. Consider signing up for vendor or government cybersecurity advisories.

» **Social media and forums.** Follow cybersecurity experts on social media platforms or join online forums. This can be an easy way to learn about new threats before they reach mainstream news.

BEST PRACTICES

» **Use official sources.** Stick to reliable outlets and experts who are verified within the cybersecurity community.

» **Practice skepticism.** Not all sources are created equal. If a security "tip" sounds dubious, cross-reference it with reputable sources.

» **Keep up-to-date.** Staying informed empowers you to take timely actions, like installing patches immediately or warning family members about a new scam.

Regularly Review Your Security Settings

Even the most meticulous security setup can become outdated if you don't periodically review and update it. As you add or replace devices, install new apps, or make changes to your network, some settings might need revisiting.

WHY THIS MATTERS

Over time, apps can gain permissions that you may have forgotten about, devices get replaced, and software updates can alter settings. A regular review ensures that everything stays adequately locked down.

STEP-BY-STEP GUIDE

» **Set a review schedule.** Decide on a monthly or quarterly check-in. Mark it on your calendar so you don't forget.

» **Check device settings.** Review each device, including computers, smartphones, routers, and Internet of Things (IoT) devices. Confirm that encryption, firewalls, and automatic updates are still enabled.

» **Review passwords.** Ensure your passwords remain strong and unique. If you suspect any were compromised, update them immediately.

» **Connected accounts.** Remove any old accounts you no longer use. Deactivate or delete unnecessary app permissions and integrations.

BEST PRACTICES

» **Use a checklist.** Jot down all the settings and accounts to review; this helps ensure you don't miss any.

» **Stay organized.** Keeping records, such as making a note in your password manager, can streamline future reviews.

» **Review periodically.** Regular reviews are like spring cleaning for your digital life: out with the old risks and in with the secure and updated configurations.

Teach Your Family

Cybersecurity is a team effort, especially if you live with others who share your network or devices. A single unsafe click can compromise the most substantial technical safeguards, so everyone in your household must understand the basics of online safety.

WHY THIS MATTERS

Children and less tech-savvy family members can inadvertently expose your entire network to malware or phishing attacks. By teaching them foundational security principles, you reduce the risk of accidental breaches.

STEP-BY-STEP GUIDE

» **Establish House rules.** Make sure everyone knows there are "rules" they must follow and take seriously.

» **Create simple guidelines.** Avoid clicking on suspicious links, refrain from downloading unknown apps, and always verify if you're unsure about an email or message.

» **Use real examples.** Show them examples of phishing emails or scam messages so they learn how to spot telltale signs, such as misspellings, strange links, or urgent threats.

» **Promote open communication.** Encourage everyone to speak up if they notice something unusual, such as a suspicious pop-up or an odd email. It's better to ask questions than to fall victim to an attack.

BEST PRACTICES

» **Kids' devices.** Set parental controls and help them understand why certain websites or apps are off-limits.

» **Frequent refreshers.** Revisit the rules and demonstrate new threats periodically; threats evolve, so knowledge must too.

» **Cooperation.** When your entire household is on the same page about security, you build a collective shield against cyber threats.

Security is a Journey, Not a Destination

The final and most crucial point to remember is that security is never truly complete. As technology advances and criminals develop new methods, your defenses must also evolve. Be resolved to go from chaos to order.

WHY THIS MATTERS

Becoming complacent can turn you into an easy target. Regular maintenance, continuous learning, and adaptation keep you one step ahead of emerging threats.

STEP-BY-STEP GUIDE

» **Adopt a growth mindset.** Recognize that there's always more to learn. Stay curious about new security tools and tactics.

» **Make small, steady improvements.** Don't overwhelm yourself by trying to do everything at once. A handful of well-implemented security measures is better than a scattershot approach.

» **Be proactive, not reactive.** If you discover a vulnerability or learn about a new type of scam, update your habits and systems immediately rather than waiting for a problem to occur.

BEST PRACTICES

» **Stay flexible.** What works today may be outdated tomorrow. Remain open to adjusting your setup if a better solution becomes available.

» **Celebrate small wins.** Updating a single password or teaching someone about phishing might seem trivial, but every step adds up to a more secure environment.

» **Pay attention.** Recognizing that security requires ongoing attention helps you maintain strong defenses even as the digital landscape evolves around you.

Moving Forward

Cybersecurity isn't just about installing one or two protective measures and hoping for the best; it's a constantly evolving discipline that requires awareness, adaptability, and a willingness to learn. Through the previous chapters, you've gained insights into securing your home network, safeguarding your IoT devices, protecting your passwords, and much more.

By following the recommendations in this guide, you can significantly improve your personal and household security posture, even if you're not a technology expert. Whenever you're unsure how to configure something, consult official manuals, documentation, or seek professional advice. Stay safe, stay updated, and keep building on these fundamentals to maintain a robust security shield for you and your loved ones.

Core Takeaways

Next, I'll recap the core ideas presented in this book, offer a few additional advanced suggestions, and provide resources to help you deepen your knowledge over time.

START WITH THE BASICS:

» **Home Wi-Fi security.** Use strong encryption (WPA2/WPA3), change default router passwords, and monitor connected devices. If you have a more advanced router or firewall, set up separate Virtual Local Area Networks (VLANs) for different device types, such as one for work devices and another for IoT devices. This extra layer of separation can contain breaches should one network segment become compromised.

» **Password management.** Rely on unique, complex passwords and consider using a reputable password manager.

» **MFA/2FA.** Enable and use multi- or two-factor authentication tools wherever they are allowed. This is one more hurdle for a hacker to access your account, even if they know your login credentials.

» **Safe application usage.** Download apps only from trusted sources and frequently review their permissions.

» **Data protection.** Back up your data regularly (remember the 3-2-1 rule), encrypt sensitive files, and delete obsolete accounts.

» **Encryption.** It's free with your device. Make sure you are using it. If you need a more robust solution or want to encrypt data sent via emails or texts, consider finding a suitable third-party solution to supplement your operating system's security features.

» **Patching and updates.** Keep software current to close known vulnerabilities and retire devices that no longer receive updates.

» **Virus protection.** Install and use a reputable antivirus program.

STAY VIGILANT:

» **Phishing and social engineering.** Always verify suspicious emails, links, and phone calls.

» **Smartphone security.** Lock your device with a strong PIN or biometric lock and avoid using public Wi-Fi whenever possible.

» **IoT best practices.** Change default credentials on smart devices and disable unnecessary features. If possible, separate these devices from your primary network.

ONGOING MAINTENANCE:

» **Regular check-ins.** Review device settings, update passwords as necessary, and verify that backups are functioning correctly.

» **Educate the household.** Ensure that everyone in your family is aware of phishing scams, password hygiene, and safe online practices.

Additional Steps to Consider

While the chapters covered many foundational topics, there are more advanced or supplemental measures you can explore:

» **Hardware security keys.** Beyond two-factor authentication apps, consider hardware-based security keys (for example, FIDO2-compatible keys) for critical accounts. These physical keys offer a robust defense against phishing attacks, as attackers would need to possess the physical device to gain access.

» **Intrusion detection systems.** If you're comfortable with more technical setups, a home Intrusion Detection/Prevention System (IDS/IPS) can alert you to suspicious network activity. This can be especially useful for individuals with extensive home networks or multiple Internet of Things (IoT) devices.

» **VPN for home and travel.** Using a virtual private network (VPN) at home can help protect your outbound traffic from certain types of monitoring. When traveling, a VPN is invaluable for securing your connection on public Wi-Fi networks, including those in hotels.

» **Zero Trust principles.** "Zero Trust" is a security model where nothing is trusted by default, not even devices or users within a supposedly trusted network. While often discussed in corporate settings, certain aspects, such as always requiring re-authentication for sensitive tasks, can be adapted at home to reduce the risk of compromised devices or accounts.

» **Physical security and device disposal.** Never discard an old device without first wiping it clean or performing a factory reset, at the very least. Wipe or destroy the storage drives of old devices before recycling or disposing of them. If you have a lot of sensitive information on the device, you may want to consider having it professionally wiped. Consider physically securing devices such as laptops with locks or in locked cabinets if they contain particularly sensitive data.

Trusted Sources and Communities

To stay current and informed, follow reputable organizations and IT professionals who consistently share valuable insights and practical advice. Here are some good starting points:

PROFESSIONAL ORGANIZATIONS

» **The SANS Institute** (sans.org). Offers free security awareness materials and newsletters.

» **Center for Internet Security** (cisecurity.org). CIS publishes configuration benchmarks and security guidelines.

» **Cybersecurity and Infrastructure Security Agency** (cisa.gov). CISA is a U.S. government agency that issues alerts and advisories on vulnerabilities and scams.

BLOGS AND TECH OUTLETS

» **Reputable tech news sites.** Monitor widely recognized technology news outlets for coverage of newly discovered threats.

» **Independent cybersecurity blogs.** Security researchers and experts often run personal blogs where they share in-depth analyses of threats and vulnerabilities.

CORPORATE RESOURCES

» **Official vendor security pages.** Companies that develop operating systems, such as Apple, Microsoft, and Google, regularly post security updates, patch notes, and best practices.

» **Device manufacturer documentation.** For routers, IoT devices, and automotive software, consult the official websites or manuals for updates, patches, and configuration tips.

LOCAL COMMUNITY AND EDUCATIONAL OPPORTUNITIES

» **Workshops and seminars.** Check local libraries, community centers, or adult education programs for basic digital security classes.

» **Online courses.** Some free platforms and Massive Open Online Courses (MOOCs) teach introductory and advanced cybersecurity topics, helping you stay knowledgeable and confident.

Progress Over Perfection

As you integrate these practices into your daily routine, remember that the goal is not to achieve "perfect" security; that doesn't exist. Instead, it's to layer (*is this the last one?*) multiple defenses that collectively make it exceedingly difficult for attackers to breach your household. Think of it like securing your home: you lock your doors, install an alarm system, and perhaps add motion-sensor lights, all while remaining cautious about who you let in.

Staying safe in the digital world requires the same mindset:

» Layer your defenses.

» Remain aware of emerging threats.

» Regularly refine and update your security measures.

By continuing to educate yourself and adapt to changing threats, you'll foster a culture of security that keeps you, your family, and your data protected for the long haul. Always remember that even minor improvements in your security posture can have a significant impact on deterring would-be intruders. Stay alert, stay curious, and carry forward the lessons from this guide to cultivate a resilient digital life. ‡

Resources

RATHER THAN TRYING TO CRAM SHRUNKEN versions of full-sized documents onto these pages, which wouldn't be very useful for you, I am providing a link to our website where you can download all the listed resources. There is no additional charge for these, and you won't be asked to sign up for a newsletter or pay membership fees to get them.

However, there is some legal language you should be aware of, and it is:

Copyright Notice and Usage Policy

Simply put, you are free to use these resources for personal purposes. However, you can't sell, use, or alter them and claim them as your own. If you require or wish to have that type of access, we can create licensing agreements with you.

The link to the downloads is: https://roguetechservices.com/downloads

Please check the site periodically, as I will continue to update and refine these resources.

WHY THIS MATTERS

- » **Organization.** When you're juggling multiple devices, accounts, and best practices, it's easy for something to slip through the cracks.
- » **Consistency.** Regularly reviewing a checklist helps you keep your

defenses up to date, especially if you revisit it every few months or after acquiring new devices.

» **Simplicity.** A clear, itemized list transforms vague "to-dos" into concrete tasks you can tackle one by one.

Here is a list of the checklists, infographics, and spreadsheets that will help you implement and verify the security measures in this book. Whether you're a busy professional juggling work accounts, a parent safeguarding your family's devices, or simply someone looking to protect your personal information, these documents help ensure you haven't missed any critical steps.

AUTHENTICATION PROS AND CONS

An infographic explaining the pros and cons of different authentication methods.

BEATING THE DEAD HORSE

An infographic that can also be used as a poster to remind you (and others) of the importance of getting secure and staying that way.

DEVICE INVENTORY SPREADSHEET

Tracking your devices, their configurations, and essential details helps you stay organized and quickly reference critical info if something goes wrong. This is a simple spreadsheet you can easily adapt and update for every device and application you own, including computers, phones, tablets, IoT devices, and software licenses.

ENCRYPTION VS. PLAIN TEXT

An infographic with an example of the difference between encrypted text and plain text.

IDENTITY THEFT PREVENTION CHECKLIST

These are steps you can take to help prevent becoming a victim of identity theft. This list also includes some warning signs that you should be aware of, which may be red flags that something is already going wrong.

IDENTITY THEFT RECOVERY CHECKLIST

If you suspect or confirm you're a victim of identity theft, act fast. This is a list of things you should address immediately if you suspect fraud on any of your accounts.

IDENTITY THEFT WARNING SIGNS

An infographic including warning signs that may be red flags that something is already going wrong.

PERSONAL SECURITY CHECKLIST

This is a high-level security checklist, divided into key categories that align with the topics in this book.

PHYSICAL SECURITY CHECKLIST

This checklist addresses physical access, cameras, door locks, IoT devices, and more.

VPN VS. VLAN

An infographic explaining the differences between the two network types.

VPN VS. VLAN-2

A different version of the infographic explaining the differences between the two.

WORK SECURITY CHECKLIST

A checklist to help you separate work and personal devices and data.

PUTTING IT ALL TOGETHER

Checklists and inventories are the practical glue that keeps all your security practices aligned. With a few structured lists covering personal accounts, physical devices, work environments, identity theft precautions, and recovery steps, you can systematically verify that you've covered every angle. Remember:

» **Update often.** As you acquire new devices or software, update your inventory accordingly.

» **Review periodically.** At least once a quarter, review your checklists to confirm that nothing has slipped through the cracks.

» **Adapt to your life.** Feel free to add more columns, reorder tasks, or create unique checklists for family members or coworkers.

Now that you have these comprehensive lists, you're well-equipped to put all the knowledge from my *Cyber Crime Defense Guide* into action, securing not just your digital environment but also offering peace of mind in a rapidly changing tech world. ⁑

Technical Terms

THIS GLOSSARY BRIEFLY REFERENCES THE TECHNICAL terms and concepts I discuss in the book. You don't need to memorize them; just scan them so they're familiar when they come up later.

» **Authentication.** A way to prove you are who you say you are when accessing a device or account. This often involves something you know (like a password), something you have (like your phone), or something you are (like a fingerprint).

» **Antivirus (AV).** A program that helps protect your devices by finding and removing harmful software (often called malware).

» **Biometrics.** Security methods that use your unique physical traits like fingerprints or facial recognition to confirm your identity.

» **Cookies.** Small files saved by websites on your browser. Some remember your preferences (like staying logged in), while others track your online behavior for advertising.

» **Data Encryption.** A method of scrambling your information so that only authorized people (or systems) can read it.

» **Data Breach.** When private information like passwords, credit card numbers, or medical records is exposed or stolen without permission.

» **Firewall.** A digital gatekeeper that blocks or allows internet traffic to protect your devices and network from threats.

» **Identity Theft.** When someone uses your personal information, like your Social Security number or bank details, to impersonate you and commit fraud.

» **Internet of Things (IoT).** Everyday items, such as thermostats, lights, cameras, or refrigerators, that connect to the internet to share data or be controlled remotely.

» **Malware.** A broad term for any software designed to damage, spy on, or take control of your device or data without your consent.

» **Multi-factor Authentication (MFA).** A security feature that asks for more than one type of proof before letting you into an account, such as entering a password and then confirming a code sent to your phone.

» **Operating System (OS).** The software that runs your computer or device. For example, Windows (on most PCs) and macOS (on Apple computers) are operating systems.

» **Passkey.** A newer, password-free login method that uses your device and a built-in security system to sign you in and is usually more secure than regular passwords.

» **Phishing.** A scam in which someone tries to trick you, often with a fake email or website, into handing over personal information like passwords or bank details.

» **Ransomware.** A type of malicious software that encrypts and locks your files and demands money to unlock them.

» **Remote Management.** A way for IT professionals to access and control a device or network from another location, it is often used for support or security.

» **Two-factor Aauthentication (2FA).** For example, a simpler version of MFA is entering your password and then typing a code texted to your phone.

» **Virtual Local Area Network (VLAN).** A way to split your home or office network into separate sections to improve security, for example, keeping your kids' devices isolated from your work computer.

» **Virtual Private Network (VPN).** A service that hides your online activity and location by creating a secure, encrypted connection and is beneficial when using public Wi-Fi.

» **Wi-Fi Encryption.** Security settings (like WPA2 or WPA3) that protect your home wireless network and make it harder for outsiders to spy on your internet traffic. ⁑

About the Author

WITH MORE THAN THREE DECADES OF experience in the ever-evolving world of technology, Edward Grassia combines a wealth of knowledge in both IT and cybersecurity. His journey began with manufacturing and robotics, where he honed a deep respect for safety and reliability. He worked for several years for major telecommunications companies during the transition from dial-up to DSL. Transitioning to K–12 education, Ed dedicated twenty-two years to protecting students and school districts from emerging digital threats.

Ed has managed large networks serving over 64,000 students and has been at the forefront of safeguarding educational institutions. In addition to configuring content filters and monitoring network traffic, he has created and led organizational security awareness trainings, gaining invaluable insights into the unique challenges children face online.

An expert in risk assessment and incident response, Ed has led teams to fend off phishing campaigns and ransomware attacks, drawing on hands-on experience with security tools and systems configuration. His passion for balancing innovation and security reflects a lifelong commitment to protecting our digital future. In the *Cyber Crime Defense Guide*, Ed shares his deep understanding of all things tech, empowering others to navigate the complexities of technology with confidence and creativity. ⁑

www.ingramcontent.com/pod-product-compliance
Lightning Source LLC
Chambersburg PA
CBHW071546200326
41519CB00021BB/6628